recent advances in phytochemistry

volume 15

The Phytochemistry of Cell Recognition and Cell Surface Interactions

RECENT ADVANCES IN PHYTOCHEMISTRY

Proceedings of the Phytochemical Society of North America

Recent Volumes in the Series

A Continuation Order Plan is available for this series. A continuation order will bring delivery
of each new volume immediately upon publication. Volumes are billed only upon actual ship-
ment. For further information please contact the publisher.

recent advances in phytochemistry

volume 15

The Phytochemistry of Cell Recognition and Cell Surface Interactions

Edited by
Frank A. Loewus
and
Clarence A. Ryan
Institute of Biological Chemistry
Washington State University
Pullman, Washington

PLENUM PRESS • NEW YORK AND LONDON

Library of Congress Cataloging in Publication Data

Main entry under title:

The Phytochemistry of cell recognition and cell surface interactions.

(Recent advances in phytochemistry; v. 15)
"Proceedings of the first joint meeting of the Phytochemical Society of North America and the American Society of Plant Physiologists, held August 4-7, 1980, at Washington State University, Pullman, Washington" — T.p. verso.
Includes bibliographies and index.
1. Plant cytochemistry—Congresses. 2. Cellular recognition—Congresses. 3. Cell interaction—Congresses. 4. Plant cell walls—Congresses. I. Loewus, Frank-Abel 1919-
. II. Ryan, Clarence A. III. Phytochemical Society of North America. IV. American Society of Plant Physiologists. V. Series.

| QK861.R38 vol. 15 [QK725] | 581.19'2s | 81-10558 |
| ISBN 0-306-40758-2 | [581.87'5] | AACR2 |

Proceedings of the First Joint Meeting of the Phytochemical Society of North America and the American Society of Plant Physiologists held August 4–7, 1980, at Washington State University, Pullman, Washington.

©1981 Plenum Press, New York
A Division of Plenum Publishing Corporation
233 Spring Street, New York, N.Y. 10013

Sweet cherry (Prunus avium L.) Blossoms

Here, the process of recognition and interaction between
pollen and pistil is expressed as growth of pollen tubes.
Self-pollinations are incompatible, as are cross pollinations
between varieties within the same S-genotype. In both incompatible and compatible pollinations, the initial events are
presumably identical but only pollen tubes from compatible
crosses continue to grow through the style until the ovary
is reached. Imcompatible crosses result in arrested growth
of pollen tubes within the style. See Chapter 8 by Clarke
and Gleeson.

Photograph by Arden Literal, WSU, 1981

PREFACE

The biological significance of carbohydrates in glycosylated biopolymers emerged from studies on viruses, microbial
cells and animal tissues. Plant-related processes, a relative newcomer to this area of research, now offer chal-
lenging questions as regards the roles of glycosyl-con-
jugates and carbohydrate-binding proteins in such broadly
based topics as pollination, fertilization, symbiosis
(including nitrogen fixation), the chemical basis of
morphogenesis, and the broad area of plant protection.
While the impressive accomplishments on model systems,
membrane-bound processes, receptor site biochemistry, and
cell surface interactions fill numerous reports, reviews,
and books, most of these involve biological systems other
than plants. A real need exists for the present volume in
which cell recognition and cell surface interactions as
related to plants are examined.

Contributions to this volume may be sorted into three
catagories: first an overview of the structures and properties of glycoconjugates, then a closer look at specific
systems in terms of biological function, and finally,
selected examples of cell recognition and cell surface
interactions as encountered in biology. To introduce the
general subject, Alan Elbein reviews the structure and biosynthesis of certain glycoconjugates and examines the biochemical basis of adherence between bacteria and eucaryotic
cells. Irwin Goldstein examines the properties of plant-
derived lectins, in particular a group of lectins from
Bandeiraea simplicifolia. The roles of several biologi-
cally-active complex carbohydrates in plant-related host-
pest relationships are examined by Peter Albersheim and
his colleagues. An in depth analysis of structural features
found in exocellular and membrane-bound glycoconjugates of
Penicillium charlesii is provided by John Gander and Cynthia
Laybourne. Glycosidase activity accompanying phytohema-
aglutinin properties of plant-derived carbohydrate-binding
proteins is described by Leland Shannon and Charles Hankins.
Robert Brown and W. C. Kimmins review their work on extraction and properties of two characteristic glycoproteins
from Phaseolus vulgaris, the so-called hydroxyproline-rich
and hydroxyproline-poor glycoproteins.

Examples of cell recognition and cell surface interactions are drawn from five biological systems. Philip
Larkin discusses plant protoplast agglutination and immobilization. Mariamne Whatley and Luis Sequeira review
the process of bacterial attachment to plant cell walls,
specifically the Agrobacterium tumefaciens, Pseudomonas
spp., Xanthomonas and Erwinia, and Rhizobia interactions.
Molecular events associated with pollen-stigma interactions,
including the immunochemistry of these events, are presented by Adrienne Clarke and Paul Gleeson. Daniel Janzen
explores the role of lectins in plant-herbivore interactions. In the final chapter, Daniel McMahon looks at
lectins as determinants for cell surface glycoconjugates
of slime mold.

The occasion of this Symposium was the first joint
meeting of the Phytochemical Society of North America and
the American Society of Plant Physiologists (in conjunction with the Western Section of the latter Society). It
was held August 4-7, 1980 at Washington State University,
Pullman, Washington. Symposium organizers included Tsune
Kosuge (Univ. Calif., Davis) and Rodeny Croteau (Washington
State Univ., Pullman), as well as the editors of this
volume. The moderators of the Symposium were Tsune Kosuge
and Leonard Beevers (Univ. Okla., Norman).

The meeting came only 10 weeks after a cataclysmic
eruption of Mount St. Helens in western Washington. Vol-
canic ash was spread over a third of the state including
Pullman. Subsequent eruptions, though less devestating,
threatened the very existance of the meeting but plant
scientists, hardy souls that they are, challenged Vulcan
at his very doorstep. Over 1,200 attended the meeting and
were rewarded with fair skies, balmy weather and a chance to
listen to the eleven outstanding papers found in this volume.

The organizers and editors wish to thank all contribu-
tors for their efforts and prompt submission of manuscripts.
Particular thanks goes to the National Science Foundation
for a grant in support of travel for our participants and
to the Graduate School, Washington State University and
the Western Section, ASPP for generous funds. It was
these sources that brought us to our goal.

Finally, acknowledgement must be made to Jane Bower
in the Word Processing Center, College of Agriculture,
Washington State University, under the direction of Doris
Birch who provided camera-ready copy of excellent quality.

CONTENTS

Chapter One

THE STRUCTURE AND BIOSYNTHESIS OF LIPOPOLYSACCHARIDES AND GLYCOPROTEINS

ALAN D. ELBEIN

Department of Biochemistry
University of Texas Health Science Center
San Antonio, Texas 78284

INTRODUCTION

The term recognition is defined by Webster as perceiving something clearly or perceiving something previously known. In that context, living cells are able to recognize various things in their environment, and this recognition must be one of the initial steps of many different cellular events.

Cell recognition can be visualized at several levels of complexity as shown schematically in Figure 1. At what might be considered the simplest level, cells recognize and interact with many types of molecules in their surroundings as depicted by the first example. These molecules, which are called chemical signals or mediators, may be as simple in structure as an amino acid or as complex as a protein. Examples of some chemical signals are hormones such as insulin, steroid hormones, auxins, gibberellins and so on. Or neurotransmitters, such as acetylcholine, or even compounds such as cholera toxin may be chemical signals.

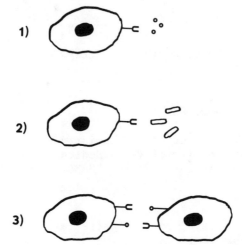

Figure 1. Model showing various types of recognition. In
1, cells recognize and interact with various small molecules
of their environment such as hormones. In 2, eucaryotic
(host) cells interact with bacteria or other microorganisms.
In 3, eucaryotic cells adhere to each other.

These chemical signals interact with cellular structures
called receptors, and this interaction must be initiated by
some sort of recognition by the receptor for its specific
mediator. Several of these systems have been studied in
depth in animal cells and a number of receptor molecules
have been isolated.

The highest level of complexity is probably the interaction between two different eucaryotic cells as is shown
in example 3. Cell adhesion and cell communication result
from this kind of interaction, and these are obviously important events in development, fertilization as well as other
cell functions. Thus, it is clear that cells have the capacity to recognize like and unlike.[1,2] An excellent example
of this level of recognition in plant systems is the pollen
grain-stigma interaction which will be discussed later in
this symposium by Dr. Clarke (See Chapter 9).

At an intermediate level of complexity, at least in
terms of the cells involved, microorganisms interact with

Table 1.   Some examples of plant-bacterial interactions
_____

    I.   Host-Parasite Interactions
           Pseudomonas solanacearum - potato, tobacco
           Agrobacterium tumefaciens - moss, beans

   II.   Host-Symbiont Interactions
           Rhizobium species - legumes

_____

plant and animal cells as shown in example 2 of Figure 1.
There have been numerous reports on the adherence of bacteria to eucaryotic cells, and these kinds of interactions
have important implications in a variety of diseases[3,4] as
well as in symbiotic associations. In this discussion, I
would like to concentrate on the interaction between bacteria
and eucaryotic cells.

    Perhaps a good starting place is a quotation from
Burnett:[5]

> "All those positive recognitions between cells
> are readily interpreted as arising from specific
> union, reversible or irreversible, between
> chemical groupings on the surfaces of inter-
> acting cells."

As pointed out here, the interaction between cells and
various ligands must involve the union between various
chemical groups at the surfaces of the interacting species.
So the question to be answered is what kinds of macromolecules are present on the surfaces of bacterial cells and
what sorts of chemical groups do they interact with on host
cell surfaces.

    In plants there are several types of bacterial inter-
actions that have been studied in some depth as indicated
in Table 1. For example, there are host-parasite interactions as exemplified by Pseudomonas solanacearum, a pathogen of tobacco and potato[6] or by Agrobacterium tumefaciens,
an organism that causes crown gall tumor in various plants.[4—7]
Some of these systems will be discussed later in this symposium by Drs. Whatley and Sequeira. In addition, there
are host-symbiotic associations as shown by the interaction

of various Rhizobium species with a number of leguminous
plants.[8,9,10]

All the associations just mentioned involve gram-
negative bacteria, and in all of these cases the bacterial
cell wall lipopolysaccharide has been implicated in the
adherence process. Thus, the first part of this discussion
will involve the structure and biosynthesis of bacterial
lipopolysaccharide. The other component of the recognition
system, that is the host component, has been suggested to
involve lectins, which are carbohydrate-binding proteins.
Since many lectins are glycoproteins, the second part of
this discussion will consider the structure and mechanism of
biosynthesis of the asparagine-linked glycoproteins. And
finally, in the last part of this discussion, I will briefly
present some data on one system of bacterial adherence to
show one example where a glycoprotein has definitely been
implicated as a receptor molecule.

STRUCTURE AND BIOSYNTHESIS OF BACTERIAL LIPOPOLYSACCHARIDES

In terms of the bacterial lipopolysaccharides, Figure 2
shows the orientation of these molecules in the bacterial
cell wall.[11] In gram-negative bacteria, the cell-envelope
is composed of the cytoplasmic membrane, a layer of peptidoglycan and an outer membrane. The cytoplasmic membrane, like

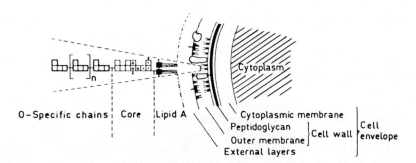

Figure 2. The organization of the gram-negative bacterial
cell envelope. The location and organization of bacterial
lipopolysaccharide is shown as an enlarged segment of the
cell wall.

membranes from other types of cells, contains phospholipid
and protein and these components are arranged as a phospholipid bilayer interspersed with integral proteins. On the
outside of this cytoplasmic layer is a rigid layer of peptidoglycan. The peptidoglycan is a continuous network of
carbohydrate chains that are crosslinked by peptide bridges,
and this layer covers the whole surface of the cell in one
giant macromolecule. This layer may be thought of as being
analogous to a fishnet surrounding a balloon and is thought
to give strength and rigidity to the cell.

External to the peptidoglycan is the outer membrane,
which contains phospholipid, lipoprotein and lipopolysaccharide. The lipoprotein portion of this membrane is presumably covalently attached to the peptidoglycan and holds
this outer membrane in place. The lipopolysaccharides are
oriented at the external surface of the outer membrane in the
arrangement shown in Figure 2. The lipid A portion of the
molecule, which is quite hydrophobic, is oriented in the membrane and is probably associated with the phospholipids and
the hydrophobic lipoprotein. On the other hand, the polysaccharide portion of the molecule composed of core polysaccharide and O-antigen side chains is at the surface and interfaces with the environment. In fact, it has been shown that
the O-antigen portion is the most exposed to external agents
and is the most antigenic portion of the molecule. Lipopolysaccharides have also been shown to be the receptor sites for
a number of bacterial viruses. Thus, the lipopolysaccharides
are likely candidates to be involved in recognition and attachment of bacteria to plant or animal cells. However, I
should point out that many gram-negative bacteria, notably
the Rhizobia, produce extracellular polysaccharides or capsules and these polymers could cover or mask the lipopolysaccharide. Space restrictions do not allow us to go into the
many diverse structures of bacterial capsules, but they should
be considered as a possible factor in bacterial adherence.

Figure 3 shows a more detailed structure of a lipopolysaccharide molecule. This particular structure is the lipopolysaccharide of Salmonella typhimurium which has been extensively studied.[12] Lipopolysaccharides from other bacteria
probably have the same general structure, but may differ in
terms of the specific sugar residues, the linkages and so on.
Some of the differences will become evident as we discuss
these structures in more detail.

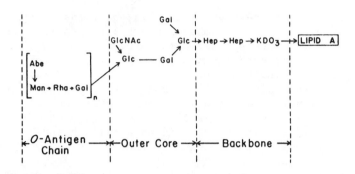

Figure 3. The structure of the lipopolysaccharide of
Salmonella typhimurium. Detailed structure of the backbone
polysaccharide, outer core, and O-antigenic side chains is
shown.

The lipopolysaccharides contain a hydrophobic core called lipid A, which is composed of glucosamine or some other
aminohexose to which fatty acids are attached in ester bonds.
Other sugars have also been found associated with lipid A in
photosynthetic bacteria. Attached to lipid A is a backbone
region of 2-keto-3-deoxyoctonic acid (KDO) and heptose (Hep).
The configuration of the heptose apparently varies in different bacteria. Linked to the backbone is a pentasaccharide
region called the outer core which is composed of the sugars
glucose, galactose and N-acetylglucosamine (GlcNAc).

From a physiological standpoint, the most important
region of the lipopolysaccharide may be the O-antigen side
chain, which is involved in many reactions of these molecules.
This is also the region that shows the greatest variation,
both within closely related species and in diverse groups of
bacteria. The O-antigen is attached to the glucose moiety
of the outer core and is usually a long polymer made up of a
number of repeating units. In Salmonella typhimurium, the
repeating unit is the tetrasaccharide shown in Figure 3, i.e.,
a repeating trisaccharide of mannose-rhamnose-galactose
with side branches of the deoxyhexose, abequose. In many
Salmonella species, this same trisaccharide repeating unit
is found, but the sugar abequose may be replaced by another
sugar, or may be completely absent. Also, in some of these

Salmonellas, the linkages between the sugars in the trisaccharide and the anomeric configurations may vary.[13] Other
organisms may have quite different O-antigenic side chains.
For example, Escherichia species are usually much simpler
than those shown here for Salmonella and contain fewer
different kinds of sugars. The important point to stress
from all of these data is that because of the enormous
diversity that is possible, the O-antigens are prime candidates for a role in recognition. Unfortunately, there is
not enough information available about the structures of the
lipopolysaccharides of Agrobacteria or Rhizobia to tell us
what their O-antigens are like, or to be able to point to
differences among species. We hope such data will be forth-
coming.

    In terms of biosynthesis of these lipopolysaccharides,
a great deal of research has been done in the laboratories
of Osborn, Rothfield, Robbins, Heath and others (see several
reviews listed on lipopolysaccharides). All of these studies
have shown that the biosynthesis involves two rather different mechanisms; in the case of the core sugars, there is a
direct transfer of sugars from their nucleotide sugar derivatives, whereas the O-antigenic side chains are synthesized via
lipid intermediates. As an example of the first case, the
core sugars glucose, galactose and GlcNAc are transferred
from their nucleotide sugars, UDP-glucose, UDP-galactose and
UDP-GlcNAc, to the lipid A backbone structure by individual
glycosyl transferases which apparently reside in the cell
envelope. Some of these glycosyl transferases have been
highly purified by Rothfield and coworkers [14, 15] and their
properties have been studied.

    Biosynthesis of the O-antigenic side chain, on the other
hand, occurs in a rather interesting and unique manner that
involves the participation of lipid carriers. In this mechanism, the sugars are transferred from nucleotide sugars to the
lipid carrier to form a lipid-linked saccharide intermediate.
The general structure of the lipid carriers in these reactions
is shown in Figure 4. The upper structure shows that the
carrier is a polyisoprenol made up of repeating isoprene
units. In the bacterial systems, the polyisoprenol contains
50 to 55 carbons, or 10 to 11 isoprene units, all of which
are unsaturated. A little later in this discussion, we will
consider the biosynthesis of glycoproteins which involve
similar types of lipid carriers that differ somewhat in

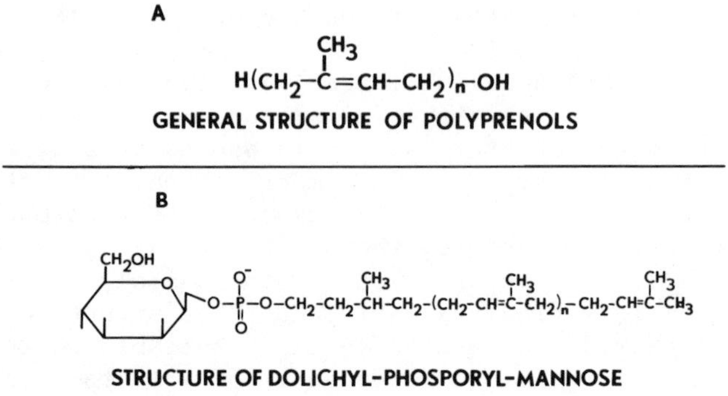

A

$$CH_3$$
$$H(CH_2-C=CH-CH_2)_n-OH$$

GENERAL STRUCTURE OF POLYPRENOLS

B

STRUCTURE OF DOLICHYL-PHOSPORYL-MANNOSE

Figure 4. Structures of carrier lipids that participate in glycosylation reactions. A shows the general structures of the polyisoprenols which are composed of repeating isoprene units. The bacterial lipids are composed of unsaturated isoprene units whereas in eucaryotic cells, the dolichols have an α-saturated isoprene unit. B shows the structure of one of the lipid-linked saccharides that participate in glycoprotein synthesis.

terms of chain length and number of unsaturations. The lower structure in Figure 4 shows one lipid-linked saccharide, dolichyl-phosphoryl-mannose, that is involved in glycoprotein biosynthesis.

Figure 5 shows the mechanism of biosynthesis of the O-antigen of Salmonella. The lipid carrier, called ACL (for antigen carrier lipid), is located in the cytoplasmic membrane of these bacteria. The individual sugars, mannose, rhamnose, and galactose, as well as any other sugars in the O-antigen, are transferred from their sugar nucleotides to the ACL to form the trisaccharide-lipid, which then becomes further elongated by the sequential addition of trisaccharide repeating units. At some undetermined stage of polymerization, this polysaccharide is transferred to the core region of the lipopolysaccharide to give a completed lipopolysaccharide molecule. Exactly how the lipopolysaccharide moves from the cytoplasmic membrane to the outer membrane is not clear at this time, nor is it clear how the degree of polymerization is determined.

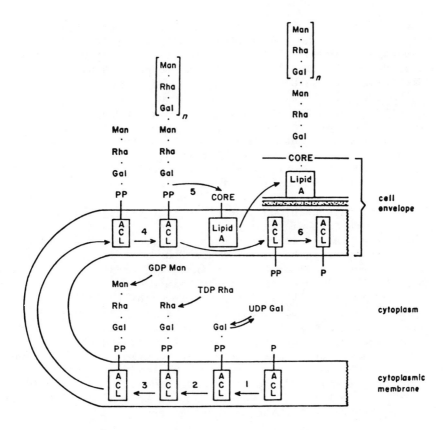

Figure 5. Mechanism of assembly of th O-antigenic side chain is shown. The antigen carrier lipid (ACL), which is located in the cytoplasmic membrane serves as the carrier for sugars which are donated from their nucleotide sugar derivatives.

STRUCTURE AND BIOSYNTHESIS OF ASPARAGINE-LINKED GLYCOPROTEINS

As indicated earlier, various workers have suggested that the plant components of these recognition systems are lectins. Many plant lectins are glycoproteins, and Figure 6 shows the structure of the oligosaccharide portion of soybean

lectin, one such glycoprotein that has been well studied.[16]
This is one of the glycoproteins that are referred to as
asparagine-linked, referring to the fact that the oligosaccharide is attached to the amide nitrogen of asparagine in a
GlcNAc-asparagine bond. These kinds of oligosaccharides,
which are called "high-mannose", because they contain large
amounts of mannose, generally contain a common core structure. This core region is a pentasaccharide composed of 2
GlcNAc residues linked in a (1→4) β-bond to which are attached
3 mannose residues. The first mannose is linked to the GlcNAc
in a (1→4) β-bond while the next 2 mannoses are attached in
an (1→3) and (1→6) α-linkage to give a branched structure.

From this point on, there may be considerable variation
in the structure of these high-mannose oligosaccharides depending on the glycoprotein being considered. For soybean
agglutinin, shown here, there are 7 or more additional mannose residues; all of them are in (1→6), (1→3) or (1→2)
α-linkages. This gives a highly branched structure. However,
other glycoproteins may differ in terms of the number of mannoses, the extent of branching or the linkages involved. In
the case of animal cells, there are additional complexities
since the asparagine-linked oligosaccharides may contain
other sugars such as galactose, sialic acid or fucose. However, in plants, so far as we know at this time, other sugars
are not present in these oligosaccharides.[17, 18] I should
point out that there is no evidence at this time to show that
the carbohydrate portion of these molecules is involved in
the receptor or recognition activity. In fact, the carbohydrate portion may be necessary in order to have the protein
transported to the proper site in the cell, or for the protein
to be inserted into the membrane, or simply for the protein
to assume the proper configuration.

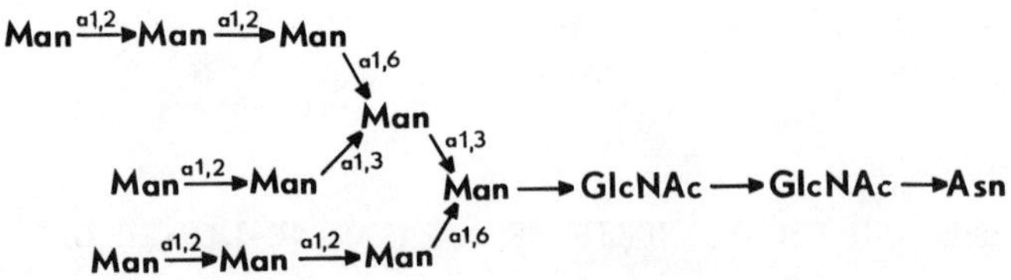

Figure 6.   Structure of the oligosaccharide portion of soybean agglutinin. The oligosaccharide is linked to protein
in a GlcNAc-asparagine bond.

As far as the formation of the oligosaccharide portion of these glycoproteins is concerned, this biosynthesis also involves the participation of lipid carriers in an analogous way to that previously discussed for lipopolysaccharides. As previously indicated in Figure 4, the lipid carrier is also a polyisoprenol, but in these glycoprotein systems, the polyisoprenol is usually a C_{100} to C_{110} lipid with the α-isoprene unit being saturated. These saturated isoprene compounds are called dolichols, and the pathway of synthesis is referred to as the dolichol pathway. Figure 5 shows the structure of dolichyl-phosphoryl-mannose, one of the lipid-linked saccharides in this pathway.

The dolichol pathway of oligosaccharide biosynthesis is presented in Figure 7.[19,20] This pathway is initiated by the transfer of GlcNAc-1-P from UDP-GlcNAc to dolichyl-P to form the first lipid-intermediate, dolichyl-pyrophosphoryl-GlcNAc. A second GlcNAc is added to form the N,N'-diacetylchitobiose-lipid, and this disaccharide is lengthened by the addition of a number of mannose residues. Some of these mannoses come directly from GDP-mannose while others come from the lipid, dolichyl-phosphoryl-mannose. This gives rise to a heptasaccharide-lipid containing 5 mannose and 2 GlcNAc residues. From this point on the further reactions of these lipids in the plant systems is not absolutely clear. In animal systems which form similar types of high-mannose oligosaccharides, there is good evidence to show that the lipid-linked oligosaccharide is further lengthened by the addition of 3 glucose and 4 mannose residues to form an oligosaccharide-lipid with the composition $Glc_3Man_9GlcNAc_2$. In the final step of this lipid carrier system, the oligosaccharide is transferred to protein while the protein is being synthesized on membrane-bound polysomes.

Following the transfer of oligosaccharide to the protein and the completion of the protein chain, the glycoprotein undergoes a number of processing or trimming reactions as outlined in Figure 8. Thus, within a short time after transfer of oligosaccharide, there is a rapid removal of all of the glucose residues by fairly specific membrane-bound glucosidases. The first one to two glucoses are apparently removed while the protein is still located in the endoplasmic reticulum, but the last glucose may be removed after the protein has been translocated to the Golgi apparatus. In the Golgi apparatus, a number of mannose residues may also be

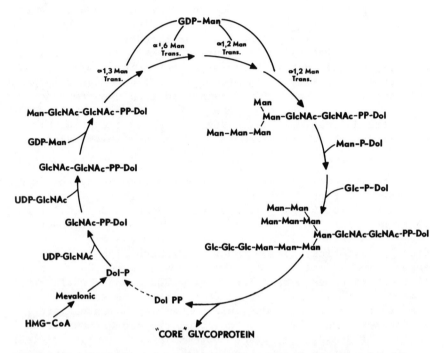

Figure 7. Pathway of biosynthesis of the oligosaccharide
chains of asparagine-linked glycoproteins as demonstrated in
animal cells. GlcNAc, mannose and glucose are added to the
lipid carrier to give an oligosaccharide $(Glc_3Man_9GlcNAc_2)$-
lipid. The oligosaccharide is then transferred to protein.

removed depending on the protein in question. In the Golgi,
a membrane-bound α-mannosidase had been identified. In the
case of animal systems, after trimming of a number of mannoses, the other sugars such as galactose and sialic acid
may be added. But those types of reactions are not known in
plant systems. So far there is nothing known about what
controls the removal of mannose residues from these proteins.
Nor is it clear what mechanism allows mannose residues to be
removed from some glycoproteins, but not others.

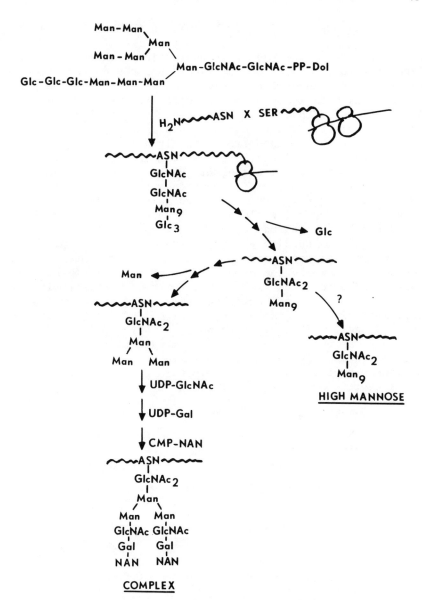

Figure 8. Reactions involved in the processing of the oligo-saccharide chains of asparagine-linked glycoproteins. After transfer of oligosaccharide to protein, all of the glucoses and a number of mannose residues are removed by membrane-bound glycosidases. In animal cells, the protein may become a "high mannose" or a "complex" type of oligosaccharide.

ROLE OF GLYCOPROTEIN IN RECOGNITION AND ADHERENCE OF BACTERIA

Now in terms of the functional role of these cell surface macromolecules, I would like to discuss one system of
bacterial adhesion which we have been studying where a glycoprotein has definitely been implicated in the recognition
and adherence of bacteria. Although this is not a system
involving plants, the techniques used in these studies and
the information obtained are certainly applicable to recognition systems in plants. This system is outlined in
Figure 9. Several years ago, Dr. Barbara Sanford observed
that upon infection of a canine kidney cell line with influenza virus, these infected cells contained receptors to which
group B streptococci would adhere. However, these bacteria
would not bind to normal uninfected kidney cells.[21] Influenza virus is a budding virus and the mature virus particle
has two coat glycoproteins; one of these glycoproteins is a
neuraminidase while the other is a hemagglutinin. When the
host cell is infected with virus, the cell machinery is
converted into the production of viral components. These
two viral glycoproteins are synthesized and inserted into
the host cell membrane. Thus when the virus particle leaves
the cell, it buds off a piece of membrane taking the viral
glycoproteins. So the preliminary results in this system
suggested that the viral coat glycoproteins were serving as
the receptors of adherence of the group B streptococci.

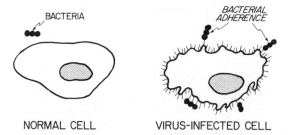

BACTERIA

BACTERIAL
ADHERENCE

NORMAL CELL   VIRUS-INFECTED CELL

Figure 9. Schematic model showing the binding of group B
streptococci to influenza virus-infected kidney cells, but
not to uninfected kidney cells. In the infected cells,
viral glycoproteins are inserted into kidney cell membrane.

In order to study this system in more detail, we developed a quantitative assay for examining bacterial adhesion. For this purpose we labeled the group B streptococci with a radioactive tag, either by growing them in ^{14}C-fructose or by reacting them with a ^{3}H-thiocyano compound which forms covalent bonds with free amino groups at the cell surface.[22] The binding of these labeled bacteria to kidney cells was then followed as shown in Figure 10. In these experiments the kidney cells were grown in plastic dishes to confluency. In some cases the confluent monolayers were infected with influenza virus and all of the cultures were allowed to incubate for an additional 24 hours. At the end of this time, the medium was removed by aspiration and the monolayers were washed well with phosphate-buffered saline. The kidney cell monolayer was then incubated for 1 hour with various amounts of ^{3}H-group B streptococci. After 1 hour, the unattached bacteria were removed by aspiration, the monolayers were washed 3 or 4 times with saline to remove free bacteria, and the cell monolayer was deteched from the plates with trypsin. The detached cells were then placed in scintillation vials

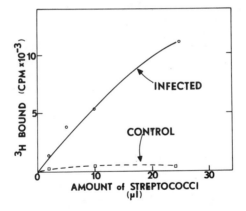

Figure 10. Radioactive assay of bacterial adhesion. ^{3}H-labeled group B streptococci were mixed in increasing amounts with either virus-infected or uninfected (control) kidney cells. After thorough washing to remove unbound bacteria, the amount ^{3}H-associated with the cell monolayer was measured.

for the determination of the amount of labeled bacteria bound
to these cells. It can be seen from Figure 10 that the group
B streptococci bound well to influenza virus-infected kidney
cells, and this binding showed saturation at high levels of
bacteria. However, no binding of group B streptococci to
uninfected cells occurred.[22]

   The binding to the virus-infected kidney cells was quite
specific for group B streptococci as shown by inhibition
studies. Thus the adherence of [3H]-labeled group B streptococci could be inhibited if the labeled bacteria were mixed
with increasing amounts of unlabeled group B streptococci.
However, unlabeled streptococci from a number of other serological groups, besides group B, had very little effect on
the adherence. These data indicate that the receptor on the
kidney cell surface is specific for some recognition site in
the group B streptococci. That this recognition site in the
group B streptococci resides in the cell wall is shown by the
experiment presented in Figure 11. As shown here, when cell
walls prepared from group B streptococci are mixed with the
kidney cell monolayers, they block the attachment of [3H]-group
B streptococci. Our interpretation of these results is that
the cell walls bind to the kidney cell receptors and tie them

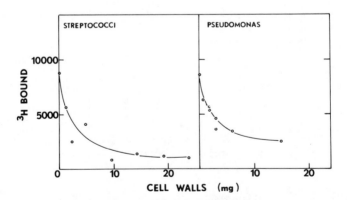

Figure 11.   Inhibition of binding of [3H] group B streptococci to virus-infected kidney cells by increasing amounts
of group B streptococcal cell walls. Various amounts of
cell walls were mixed with kidney cell monolayers and after
an incubation of 30 min, the monolayers were challenged with
[3H]-bacteria.

up so that they are not available for ginding the intact
bacteria Interestingly enough, as shown on the right hand
side of this figure, Pseudomonas cell walls also block the
adherence of group B streptococci to these cells. In other
experiments it was found that Pseudomonas aeruginosa also
binds to the kidney cells, but in this case the Pseudomonas
adhere equally well to the infected and normal cells. Apparently the receptor for these bacteria is not the viral glycoproteins, but is another receptor normally present in these
cells.

Evidence that the receptor for group B streptococci is
indeed the viral glycoproteins is shown in Figure 12. In
this experiment, various amounts of group B streptococci
were mixed with free influenza virus and this mixture was
tested for its ability to bind as compared to group B streptococci not mixed with virus. It can be seen from the
figure that in the presence of virus, fewer group B streptococci bound to the kidney cells. We assume that the reason
for this observation is that the group B streptococci bind
to influenza virus and therefore are not available to bind
to kidney cells. These studies also implicate the viral
coat glycoproteins as the streptococcal receptor.

The influenza virus coat glycoproteins have been
partially characterized in terms of their carbohydrate composition and they are asparagine-linked oligosaccharides of
both the "high mannose" and the "complex type" as shown in
Figure 13. Apparently both of these types of oligosaccharides are found in the viral hemagglutinin and the
neuraminidase.[23] As indicated earlier in this discussion,
these asparagine-linked oligosaccharides are synthesized by
means of the lipid-linked saccharide pathway discussed
earlier. There is a very useful antibiotic called tunicamycin which has been shown to block the formation of some of
the lipid-linked saccharides. More specifically, tunicamycin blocks the reaction shown in Figure 14, that is, it prevents the formation of dolichyl-pyrophosphoryl-GlcNAc.[24, 26]
Since this lipid is the first lipid intermediate in the synthetic pathway for glycoprotein synthesis if the formation of
this compound is blocked, then the lipid-linked oligosaccharides cannote be formed and the protein cannot be glycosylated.
The tunicamycin is very useful for studying the role of the
carbohydrate in the function of various glycoproteins. Since
the influenza viral glycoproteins involve the lipid-linked

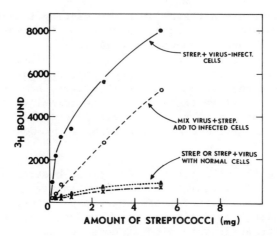

Figure 12. Effect of mixing free virus with group B
streptococci. Increasing amounts of [³H] streptococci were
mixed with free virus and this mixture was compared to group
B streptococci alone for its ability to bind to virus-
infected kidney cells.

```
                              NeuAc  NeuAc    NeuAc  NeuAc
                                |      |        |      |
                               Gal    Gal      Gal    Gal
                                |      |        |      |
 Man   Man   Man             GlcNAc GlcNAc   GlcNAc GlcNAc
    \   |   /                     \  /           \  /
    Man   Man                      Man            Man
       \ /                            \          /
       Man                             \        /
        |                               Man
      GlcNAc                             |
        |                             GlcNAc
      GlcNAc                             |
        |                             GlcNAc
       Asn                              |
       (A)                             Asn
                                       (B)
```

Figure 13. The "high mannose" (A) and "complex" (B) types of
oligosaccharides found in the influenza virus glycoproteins.
Although the detailed structures of the viral glycoproteins
are not known, they have a general structure like those
shown here.

saccharide pathway,[27] it was of interest to determine the effect of tunicamycin on the synthesis of these molecules and on bacterial adhesion.

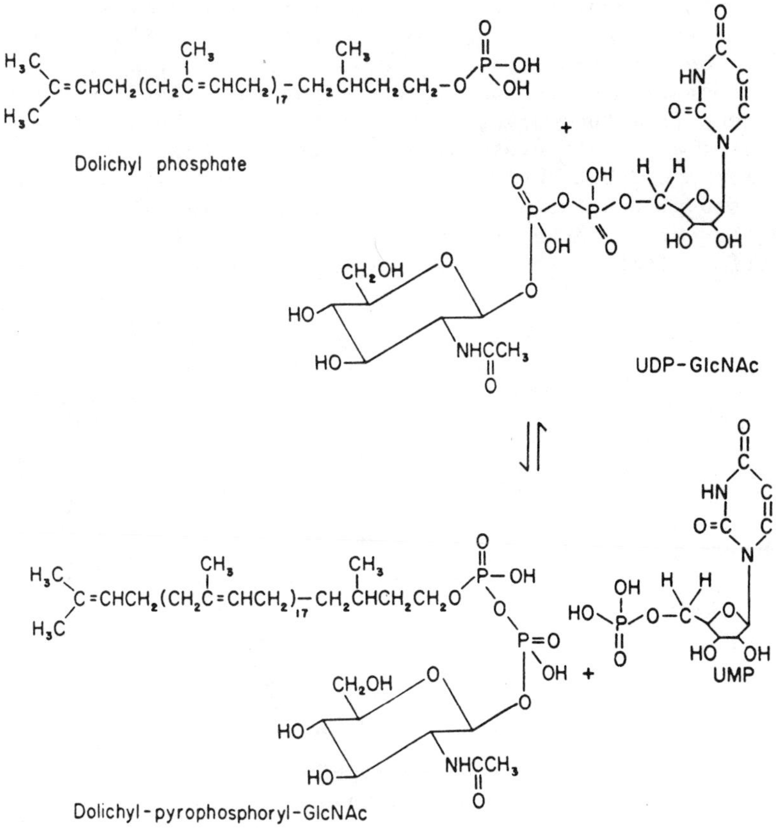

Figure 14. Tunicamycin sensitive reaction in the lipid-linked saccharide pathway. The transfer of GlcNAc-1-P from UDP-GlcNAc to form dolichyl-pyrophosphoryl-GlcNAc is the first step in the dolichol pathway.

    Figure 15 shows the effect of tunicamycin on the
synthesis of protein and on protein glycosylation in the
virus-infected kidney cells. In this experiment, kidney
cell monolayers were infected with influenza virus and after
one hour, 0.9 µg/ml of tunicamycin was added. The cells
were incubated with antibiotic for another hour in order to
allow the antibiotic to take effect and then various radioactive precursors were added. [³H]Mannose was added to
these cells as a measure of the incorporation of mannose
into protein (i.e., protein glycosylation), and [³H]leucine
to measure leucine incorporation into protein (i.e., protein
synthesis). After incubation of these cells with the labeled
precursors for 3 hours, the cells were harvested and the
incorporation of isotope into protein was measured. It can
be seen from the curves on the left that mannose was rapidly
incorporated into protein in control cultures (i.e., minus
antibiotic), but tunicamycin almost completely inhibited
protein glycosylation at about 1 µg/ml. However, as shown
by the curves on the right, this antibiotic had relatively
little effect on the incorporation of leucine into protein.

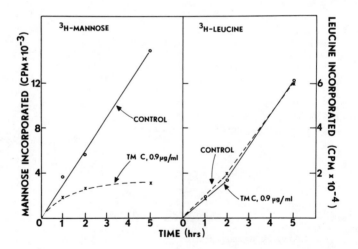

Figure 15.   Incorporation of [³H] mannose and [³H]leucine
into protein in MDCK cells.   Virus-infected cells were
incubated for 1 hour with tunicamycin and then labeled
precursors were added for 3 hours.   Control cells did not
have tunicamycin.

Since this antibiotic prevented the glycosylation of proteins, it was of interest to examine its effect on the adherence of streptococci in these virus infected kidney cell monolayers. The results of such an experiment are shown in Table 2. In this experiment, kidney cells were

Table 2. Effect of tunicamycin on influenza virus: Infection and binding of group B streptococci

Tunicamycin (µg/ml)	Hemadsorption plaque assay[a] (% of control)	Hemag-glutinin titer[b] (super-natant)	Binding of group B streptococci[c] (% of control)
0	100	1:128	100
0.078	100	1:16	71
0.156	50	1:11	6
0.32	7.7	--[d]	2
0.625	0	--	0
1.25	0	--	0
2.5	0	--	0
5.0	0	--	0
10.0	0	--	0

[a]Hemadsorption by kidney cells. The control (100%) is equal to the mean number of hemadsorption plaques on virus-infected monolayers which have not been pretreated with tunicamycin.

[b]Hemagglutinin released from cells as measured by agglutination of erythrocytes.

[c]Binding of streptococci as measured by microscopic observation. The control (100%) is equal to the mean number of bacterial adherent plaques on virus-infected monolayers which have not been pretreated with tunicamycin. This assay correlated well with the ^{3}H-binding assay done in other experiments.

[d]--, Undiluted supernatant was hemagglutination negative.

infected with virus and, after one hour, various concentrations of antibiotic were added. The cells were allowed to
incubate for 24 hours in order to allow the virus to repli-
cate. At the end of this time, the monolayers were tested
for the presence of virus particles and for their ability
to bind group B streptococci. The hemadsorption assay is a
measure of the presence of influenza-viral glycoproteins in
the kidney cell membrane while hemagglutination is a measure
of the release of mature virus particles into the medium.
The table shows that at about 0.3 µg/ml of tunicamycin, the
viral glycoproteins are no longer found in the kidney cell
surface, and there is no detectable virus in the medium.
At this level of antibiotic, the kidney cells also lack the
receptor to which the group B streptococci adhere. Thus,
these experiments in conjunction with those already discussed indicate that the viral glycoproteins that have been
synthesized in these infected cells and inserted into the
membrane are able to serve as receptor sites for the
attachment of certain bacteria. It is not known at this
time whether the carbohydrate portion of these molecules is
involved in bacterial adhesion. That is, the carbohydrate
portion of the glycoproteins may be necessary in order for
the protein to be inserted into the membrane and, in the
absence of glycosylation, the viral proteins may remain in
the cytoplasm of the cells.

SUMMARY

Cell surface macromolecules from bacteria and from the
eucaryotic host cells undoubtedly interact in specific ways
to allow binding of these cells to each other. In some
plant-microbe interactions, bacterial lipopolysaccharides
and plant lectins have been implicated in these processes.
However, bacterial cells have many different macromolecules
at their surface including various extracellular polysaccharides. These molecules could also be involved in recognition and adherence. In order to demonstrate which components are involved, they must be isolated from the respective
organism, purified to homogeneity, and tested both in vivo
and in vitro for their effect on adhesion.

REFERENCES

1. Clarke, A. and B. Knox. 1976. Cell recognition in
    flowering plants. Quart. Rev. Biol. 53:3-28.
2. Frazier, W. and L. Glaser. 1979. Surface components
    and cell recognition. Ann. Rev. Biochem. 48:491-523.
3. Gibbons, R. J. and J. van Houte. 1975. Bacterial
    adherence in oral microbial ecology. Ann. Rev. Microbiol.
    19:29-44.
4. Lippincott, B. B. and J. A. Lippincott. 1969. Bacterial
    attachment to a specific wound site as an essential stage
    in tumor initiation by Agrobacterium tumefaciens. J.
    Bacteriol. 97:620-628.
5. Burnet, F. 1971. Self-recognition in colonial marine
    forms and flowering plants in relation to evaluation of
    immunity. Nature 232:230-235.
6. Sequeira, L. 1978. Lectins and their role in host-
    pathogen specificity. Ann. Rev. Phytopathol. 16:453-481.
7. Whatley, M. H., J. S. Bodwin, B. B. Lippincott and J. A.
    Lippincott. 1976. Role for Agrobacterium cell-envelope
    lipopolysaccharide in infection site attachment. Infect.
    Immun. 13:1080-1083.
8. Sahlman, K. and G. Fahraeus. 1963. An electron micro-
    scope study of root-hair infection by Rhizobium. J. Gen.
    Microbiol. 33:425-427.
9. Dazzo, F. B. and D. H. Hubbell. 1975. Antigenic dif-
    ferences between infective and noninfective strains of
    Rhizobium trifolii. Appl. Microbiol. 30:172-177.
10. Schmidt, E. L. 1979. Initiation of plant root-microbe
    interactions. Ann. Rev. Microbiol. 33:335-376.
11. Weckesser, J. and G. Drews. 1979. Lipopolysaccharides
    of photosynthetic procaryotes. Ann. Rev. Microbiol.
    33:215-239.
12. Osborn, M. J. 1969. Structure and biosynthesis of the
    bacterial cell wall. Ann. Rev. Microbiol. 38:501-538.
13. Wright, A. and S. Kanegasaki. 1971. Molecular aspects
    of lipopolysaccharides. Physiol. Rev. 51:748-784.
14. Rothfield, L. and M. Perlman-Kothencz. 1969. Synthesis
    and assembly of bacterial membrane components - A lipo-
    polysaccharide-phospholipid- protein complex excreted
    by living bacteria. J. Mol. Biol. 44:477-492.

15. Romeo, D., A. Girard and L. Rothfield. 1970. Reconsti-
     tution of a functional membrane enzyme-system in a mono-
     molecular film 1. Formation of a mixed monolayer of
     lipopolysaccharide and phospholipid. J. Mol. Biol.
     53:475-490.
16. Lis, H. and N. Sharon. 1978. Soybean agglutinin - A
     plant glycoprotein. J. Biol. Chem. 253:3468-3476.
17. Sharon, N. and H. Lis. 1979. Comparative biochemistry
     of plant glycoproteins. Biochem. Soc. Trans. 7:783-799.
18. Kornfeld, R. and S. Kornfeld. 1976. Comparative aspects
     of glycoprotein structure. Ann. Rev. Biochem. 45:213-237.
19. Elbein, A. D. 1979. The role of lipid-linked saccharides
     in the biosynthesis of complex carbohydrates. Ann. Rev.
     Plant Physiol. 30:239-272.
20. Waechter, C. J. and W. J. Lennarz. 1976. The role of
     polyprenol-linked sugars in glycoprotein synthesis.
     Ann. Rev. Biochem. 45:95-112.
21. Sanford, B. A., A. Shelokov and M. A. Ramsey. 1978.
     Bacterial adherence to virus-infected cells: A cell
     culture model of bacterial superinfection. J. Infect.
     Dis. 137:176-181.
22. Pan, Y. T., J. W. Schmitt, B. A. Sanford and A. D.
     Elbein. 1979. Adherence of bacteria to mammalian
     cells: Inhibition by tunicamycin and streptovirudin.
     J. Bacteriol. 139:507-514.
23. Collins, J. K. and C. A. Knight. 1978. Purification
     of the influenza hemagglutinin glycoprotein and character-
     ization of its carbohydrate components. J. Virology,
     26:457-467.
24. Thacz, J. and J. Lampen. 1975. Tunicamycin inhibition
     of polyisoprenyl-N-acetyl glucosaminyl pyrophosphate
     formation in calf liver microsomes. Biochem. Biophys.
     Res. Commun. 65:248-257.
25. Ericson, M., J. Gafford and A. D. Elbein. 1977.
     Tunicamycin inhibits GlcNAc-lipid formation in plants.
     J. Biol. Chem. 252:7431-7433.
26. James, D. W., Jr. and A. D. Elbein. 1980. Effects of
     several tunicamycin-like antibiotics on glycoprotein
     biosynthesis in mung leaves and suspension-cultured
     soybean cells. Plant Physiol. 65:460-464.
27. Schwarz, R. T., J. N. Rohrschneider and M. F. G. Schmidt.
     1976. Suppression of glycoprotein formation of Semliki
     forest, influenza, and avian sarcoma virus by tunicamycin.
     J. Virology 19:782-791.

Chapter Two

PLANT DERIVED LECTINS

IRWIN J. GOLDSTEIN

Department of Biological Chemistry
University of Michigan
Ann Arbor, Michigan 48109

The term lectin [Latin, legere - to pick out, choose] was coined by W. C. Boyd[1] in order to call attention to a group of plant seed proteins which could distinguish among human blood groups. Boyd had discovered that extracts of the lima bean (Phaseolus lunatus) specifically agglutinated type A erythrocytes (Fig. 1).

Reaction of extract with cells of group:		
A	B	0
EW ++++	LH ±	BD 0
MF ++++		BR 0
DA ++++		SJ 0
JB ++++		ON 0
AL ++++		BA 0
WCB ++++		CTS 0

Figure 1. Test of lima bean extract (December 10, 1945)[2]

25

Table 1.   Examples of enzymes that act like lectins

---

1.   Crystalline α-amylase forms an insoluble complex with
     glycogen at 4°C.[4]

2.   Muscle phosphorylase a and b precipitate glycogen at
     0°C.[5]

3.   Galactose oxidase agglutinates sialidase-treated human
     erythrocytes at 0°C. At higher temperatures, the
     agglutinate disperses.[6]

4.   Mung bean α-galactosidase agglutinates trypsinized
     rabbit erythrocytes at pH 8.5 and 0 - 5°C (pH optimum
     of enzymatic activity is 6.5; at pH 8.5 activity is
     only 10%). At 6.5 and higher temperatures the clot
     rapidly dissolves.[7]

5.   Lysozyme cross-linked by treatment with glutaraldehyde
     agglutinates human erythrocytes.[8]

     Lectins are naturally occurring, carbohydrate binding
(glyco) proteins of non-immune origin which agglutinate
cells and/or precipitate complex carbohydrates (polysaccharides, glycoproteins, glycolipids).[3] Although no known
lectin has been shown to exhibit enzymatic activity, under
certain defined conditions some enzymes may act like lectins,
i.e. display lectin-like properties. Examples of some
enzymes which act like lectins are listed in Table 1.

     A major concern of many plant physiologists and bio-
chemists is the biological role which lectins may serve.
Although many hypotheses have been advanced, none has been
proved. Possible functions of plant seed lectins are
tabulated in Table 2.

     Although the function of lectins has not yet been
clarified, these fascinating substances display a host of
remarkable biological properties (Table 3) which find wide
application in many biomedical investigations (Table 4).

Table 2. Possible functions of plant lectins
_____
Sugar transport, storage and immobilization

Carbohydrases or procarbohydrases

Plant antibodies

Binding nitrogen-fixing bacteria to legumes

Involvement in germination, differentiation, maturation and
    cell division

Protection against insect and fungal predators

Involvement in cell wall metabolism
_____

Table 3. Biological properties of lectins
_____
Agglutination of cells and particles: erythrocytes, lympho-
    cytes, tumor cells, microorganisms, viruses, vesicles

Mitogenic stimulation of lymphocytes

Inhibition of fungal growth

Inhibition of activity of glycoprotein enzymes

Degranulation of mast cells

Insulin-like activity on fat cells

Cytotoxic activity toward mammalian cells

Modulation of the immune response
_____

        Plant seed lectins may be classified into a limited
number of carbohydrate binding specificity groups. Six
carbohydrate binding specificity groups together with
examples and some biological properties of each group are
tabulated in Table 5.

Table 4.   Some uses of lectins

Blood typing, the detection of "secretors" and
    polyagglutination

Detection, preliminary characterization and structural
    studies on complex carbohydrates

Isolation and purification of carbohydrate-containing
    molecules

Fractionation of cells and particles (e.g., viruses, cellular
    organelles, vesicles)

Models for carbohydrate-protein and antibody-antigen
    interactions

Identification of carbohydrate-containing, cell surface
    markers (antigens)

Mitogenic stimulation of lymphoctyes in clinical and
    immunological studies

Generation of lectin-resistant variants of eukaryotic cells
for studies of glycoprotein structure and metabolism

   Closer examination of these lectin classes allows us to
discern some common structural and phylogenetic features
(Table 6). For example, all plant lectins are multimeric
(glyco) proteins, i.e. they consist of 2 or more polypeptide
chains. With the notable exception of concanavalin A, all
seed lectins from the family Leguminosae are glycoproteins
and most are metalloproteins which lose all biological
activity when deprived of their metal ions. Furthermore,
most leguminous lectins are deficient in sulfur-containing
amino acids. Concanavalin A contains methionine but no
cysteine residues, the lima bean lectin (Phaseolus lunatus)
contains cysteine but no methionine residues and the lectin
from the red kidney bean (Phaseolus vulgaris) contains
neither cysteine nor methionine residues.

   A structural basis for some lectin classes which display
the same sugar binding specificity is suggested by recent

Table 5. Classification of lectins into a limited number of
carbohydrate-binding specificity groups.

---

1. D-Mannose (D-Glucose)-binding lectins

    jack bean, pea, lentil, broad bean lectins
    (blood group non-specific; mitogenic)

2. N-Acetyl-D-glucosamine-binding lectins

    B. simplicifolia II, jimson weed, potato, wheat germ
    lectins
            (bind to chitin; very weak or non-mitogenic;
            blood group non-specific)

3. N-Acetyl-D-galactosamine-binding lectins

    Dolichos biflorus, soy bean, lima bean, edible snail
    lectins
            (blood group A - specific; lima bean and
            polymerized soy bean lectins are mitogenic).

4. D-Galactose-binding lectins

    A. precatorius agglutinin, peanut, B. simplicifolia I,
    castor bean agglutinin
            (BS I - B_4 is blood group B-specific; A.
            precatorius is mitogenic).

5. L-Fucose-binding lectins

    Eel, L. tetragonolobus, Ulex europeus I lectins
            (blood group O-specific; weak or non-mitogenic)

6. Complex carbohydrate-binding lectins

    Red kidney bean (PHA) - a potent mitogen
    Meadow mushroom - a potent mitogen
    Horse-shoe crab agglutinin - binds sialic acid

---

reports on the structure and amino acid sequence of several
of these carbohydrate-binding (glyco) proteins. For example,
there are only two differences in the amino acid sequence of
the first 25 amino acids in the β-chain of the D-mannose-
binding pea and lentil lectins, and the N-terminal sequence
of amino acids of the α-chain of these agglutinins is nearly
identical.[9,10] The soybean and peanut agglutinins are identical in 11 of the first 25 N-terminal amino acids and the
N-terminal sequence of amino acids of the soybean agglutinin
is similarly identical at 11 positions with that of the β-
chain of the lentil lectin. Many of the amino acid differences observed in these cases could have resulted from a
single nucleotide substitution.

     Even more remarkable, the primary structure of the
entire α subunit from the lentil lectin was found to be homologous with the region between positions 72 and 121 from con
A and the amino-terminal sequence of the β-chain is homologous
to another portion of the con A molecule between position 123
and 165.[11] Similarly, Cunningham and his colleagues[12] found
that the α-chain of the lectin from the broad bean (Vicia
faba) is homologous to a region in the middle of the con A
sequence (residues 70-119), and that the β-chain is homologous to two discrete segments of con A comprising a
circular permutation of amino acid sequences.

     A further striking structural feature observed for
several lectins has been the almost identical sequence of
N-terminal amino acids in the two distinctive subunits
present in a single lectin or isolectin thereof. Thus, the
first 10 N-terminal amino acids of the two soybean lectin
subunits[13] and the first 30 amino terminal amino acids of
the two subunits of the Dolichos biflorus lectin[14], are identical, and there are considerable homologies evident between
the sequences of N-terminal amino acids of the two subunits
of the lectin (PHA) from Phaseolus vulgaris seeds.[15] The
discovery of extensive homologies between different lectins
strongly suggests a common genetic origin. Phylogenetic
trees have been constructed which reveal these relationships
and support the notion that lectins may have an important
physiological role in plants.

     As an example of how lectins are isolated, purified and
characterized and used in biological studies, I would like
to briefly discuss the lectins present in Bandeiraea

Table 6.   Lectins: Some physical-chemical properties.

	Subunits No./M_r	Glyco-protein	Metalo-protein
1. $\underline{D}$-Mannose ($\underline{D}$-Glucose)-binding lectins			
Jack bean (<u>Canavalia</u> <u>ensiformis</u>)	4/26,000	-	+
Lentil (<u>Lens</u> <u>culinaris</u>)	2a, 2β/5,900, 17,000	+	+
Pea (<u>Pisum</u> <u>sativum</u>)	2a, 2β/5,900, 17,000	+	+
Broad bean (<u>Vicia</u> <u>faba</u>)	2a, 2β/5,000, 20,000	+	+
2. N-Acetyl-$\underline{D}$-glucosamine-binding lectins			
<u>Bandeiraea</u> <u>simplicifolia</u> <u>II</u>	4/30,000	+	+
Jimson weed (<u>Datura</u> <u>stramonium</u>)	2, 4/40,000	+	-
Potato (<u>Solanum</u> <u>tuberosum</u>)	2/50,000	+	-
Wheat germ (<u>Triticum</u> <u>vulgaris</u>)	2/18,000	-	-
Gorse seed (<u>Ulex</u> <u>europeus</u> <u>II</u>)	4/30,000	+	+
Poke weed mitogen (<u>Phytolacca</u> <u>americana</u>)	1-5/19,000-31,000	+	?
3. N-Acetyl-$\underline{D}$-galactosamine-binding lectins			
Horse gram (<u>Dolichos</u> <u>biflorus</u>)	4/26,000	+	+
Soy bean (<u>Glycine</u> <u>max</u>)	4/30,000	+	+
Edible snail (<u>Helix</u> <u>pomatia</u>)	6/13,000	+	-
Lima bean (<u>Phaseolus</u> <u>lunatus</u>)	4, 8/31,000	+	+
4. $\underline{D}$-Galactose-binding lectins			
Jequirity bean (<u>Abrus</u> <u>precatorius</u>)	4/33,000	+	+
Peanut (<u>Arachis</u> <u>hypogaea</u>)	4/27,500	-	?
<u>Bandeiraea</u> <u>simplicifolia</u> <u>I</u>	4/28,500	+	+
Castor bean (<u>Ricinus</u> <u>communis</u>)	4/31,000, 33,000	+	-
5. $\underline{L}$-Fucose-binding lectins			
Eel (<u>Anguilla</u> <u>anguilla</u>)	3/40,000	<0.4%	-
Asparagus pea (<u>Lotus</u> <u>tetragonolobus</u>)	2, 4/29,000	+	?
Gorse seed (<u>Ulex</u> <u>europeus</u> <u>I</u>)	2/28,000	+	-
6. Complex carbohydrate-binding lectins			
Red kidney bean (<u>Phaseolus</u> <u>vulgaris</u>)	4/31,000	+	+
Vicia graminea	4/25,000	+	?
Meadow mushroom (<u>Agaricus</u> <u>campestris</u>)	4/16,000	+	?
Horse-shoe crab (<u>Limulus</u> <u>polyphemus</u>)	18-20/20,000	+	+

simplicifolia (more properly termed Griffonia simplicifolia)
seeds. The B. simplicifolia plant is a shrub which grows in
the tropical rain forests of West Africa, particularly in
Ghana. It reaches a height of 3 to 4 meters.

Thus far we have isolated 4 distinctly different lectins
from B. simplicifolia seeds (Table 7). These lectins differ
in their physical-chemical properties, carbohydrate binding
specificity, and immunochemical cross-reactivity. The
fourth lectin, BS IV, has recently been isolated and shown
to be active against Le^b blood group substance which contains
L-fucose as its immunodominant sugar.[16]

Bandeiraea simplicifolia I plant seed isolectins com-
prise a family of tetrameric α-D-galactopyranosyl-binding
glycoproteins composed of various combinations of two different kinds of subunits designated A and B.[17] [18] The A subunits exhibit a primary specificity for N-acetyl-α-D-galacto-
saminyl groups but also cross-react with α-D-galactosyl
groups whereas the B subunit shows a sharp specificity for
α-D-galactosyl units. Subtypes of the A and B subunits have
been demonstrated by isoelectric focusing.

The five isolectins have been purified by use of two
affinity columns. All five BS I isolectins bind to a
melibionate-Bio Gel column. The A_4 and A_3B isolectins were
displaced as purified components by increasing quanities of
N-acetyl-D-galactosamine. The B_4, B_3A and B_2A_2 components,

Table 7. Bandeiraea simplicifolia seeds contain 4 lectins

BS I   - A system of 5 α-D-galactosyl-binding
          isolectins

BS II - An N-acetyl-D-glucosamine-binding lectin

BS III - A system of 5 N-acetyl-α-D-galactosaminyl-
          binding isolectins

BS IV   - A lectin with Le^b-specificity

---

after elution from the melibionate column with methyl
α-$\underline{D}$-galactopyranoside, were resolved on a column of
insolubilized blood group A substance according to the
procedure of Murphy and Goldstein.[17] The latter affinity
column has $\underline{N}$-acetyl-α-$\underline{D}$-galactosaminyl determinant groups
which interact with the subunit A binding sites.

Physical-chemical characterizaton of the A_4 and B_4
isolectins revealed the A and B subunits to be homologous
structures with closely similar amino acid compositions
although they differ markedly in one respect: the B subunit has one methionine residue whereas the A subunit contains no methionine. Ouchterlony analysis indicated that in
addition to common structural features, each subunit contains
its own distinct antigenic determinants. The B subunit has a
molecular weight (M_r 33,000) approximately 1000 daltons
greater than the molecular weight of the A subunit (M_r 32,000).
These data combined with our observation that the isolectins
present in individual seeds are never present in equivalent
amounts but rather exhibit a skewed distribution have led us
to postulate a precursor-product relationship between the B
and A subunit.[19]

The BS I-B_4 isolectin has been labeled with fluorescein,
ferritin, colloidal gold and tritium and used as a probe for
the detection of α-$\underline{D}$-galactosyl groups in plant and animal
cells. The immobilized form of the lectin has also been
used for a one step purification of the galactomannan from
Cassia alata seeds.[20] Fluorescein-labeled BS I-B_4 (FITC-B_4)
was shown to bind to and agglutinate Ehrlich ascites tumor
cells, an observation confirmed by the detection on the
electron microscopy of ferritin-labeled B_4 bound to Ehrlich
ascites cell membranes.[21] Using immobilized BS I-B_4 as an
affinity matrix we have been able to isolate and characterize
a family of α-$\underline{D}$-galactosyl containing glycoproteins from an
Ehrlich cell membrane fraction. FITC-B_4 was also shown to
bind to murine kidney sections. Specific inhibition by methyl
α-$\underline{D}$-galactoside and abolition of binding by α-galactosidase
in addition to morphological evidence indicates the lectin
binds to α-$\underline{D}$-galactosyl groups of basement membrane.[22]

BS I-B_4 has also been shown to be cytotoxic to a variety
of animal cells and used to generate a variant clone of
B_4-resistant 3T3 murine cells.[23]

REFERENCES

1. Boyd, W. C. and E. Shapleigh. 1954. Antigenic rela-
    tions of blood group antigens as suggested by tests
    with lectins. J. Immunol. 73:226-231.
2. Boyd, W. C. 1962. Introduction to immunochemical
    specificity. John Wiley & Sons, New York pp. 1-158.
3. Goldstein, I. J., R. C. Hughes, M. Monsigny, T. Osawa
    and N. Sharon. 1980. What should be called a lectin?
    Nature 285:66.
4. Levitzki, A., J. Heller and M. Schramm. 1964. Specific
    precipitation of enzyme by its substrate: The α-
    amylasemacrodextrin complex. Biochem. Biophys. Acta
    81:101-107.
5. Selinger, Z. and M. Schramm. 1963. An insoluble com-
    plex formed by the interacton of muscle phosphorylase
    with glycogen. Biochem. Biophys. Res. Commun.
    12:208-214.
6. Horejsi, V. 1979. Galactose oxidase--An enzyme with
    lectin properties. Biochem. Biophys. Acta 577:383-388.
7. Hankins, C. N. and L. M. Shannon. 1978. The physical
    and enzymatic properties of phytohemagglutinin from
    mung beans. J. Biol. Chem. 253:7791-7797.
8. Horejsi, V. and J. Kocourek. 1974. Studies on
    Phytohemagglutinins XXI. The covalent oligomers of
    lysozyme - First case of semisynthetic hemagglutinins,
    Experientia, 30:1348-1349.
9. Foriers, A., C. Wuilmart, N. Sharon and A. D. Strosberg.
    1977. Extensive sequence homologies among lectins from
    leguminous plants, Biochem. Biophys. Res. Commun.
    75:980-986.
10. Foriers, A., E. Van Driessche, R. de Neve, L. Kanarek,
    A. D. Strosberg and C. Wuilmart. 1977. The subunit
    structure and N-terminal sequences of the α- and
    β-subunits of the lentil lectin (Lens culinaris). FEBS
    Lett. 75:237-240.
11. Foriers, A., R. de Neve, L. Kanarek and A. D. Strosberg.
    1978. Common ancestor for concanavalin A and lentil
    lectin? Proc. Natl. Acad. Sci. USA. 75:1136-1139.
12. Cunningham, V. A., J. J. Hemperly, T. P. Hopp, and G.
    M. Edelman. 1979. Favin versus concanavalin A:
    Circularly permuted amino acid sequences. Proc. Natl.
    Acad. Sci. USA. 76:3218-3222.

13. Lotan, R., R. Cacan, M. Cacan, H. Debray, W. G. Carter and N. Sharon. 1975. On the presence of two types of subunit in soybean agglutinin. FEBS Lett. 57:100-103.
14. Etzler, M. E., C. F. Talbot and P. R. Ziaya. 1977. NH₂-Terminal sequences of the subunits of Dolichos Biflorus lectin. FEBS Lett. 82:39-41.
15. Miller, J. B., R. Hsu, R. Heinrikson and S. Yachnin. 1975. Extensive homology between the subunits of the phytohemagglutinin mitogenic proteins derived from Phaseolus Vulgaris. Proc. Nat. Acad. Sci. USA. 72:1388-1391.
16. Shibata, S. and I. J. Goldstein. 1980. Manuscript in preparation.
17. Murphy, L. A. and I. J. Goldstein. 1977. Five α-D-galactopyranosyl-binding isolectins from Bandeiraea simplicifolia seeds. J. Biol. Chem. 252:4739-4742.
18. Murphy, L. A. and I. J. Goldstein. 1979. Physical-chemical characterization and carbohydrate-binding activity of the A and B subunits of the Bandeiraea simplicifolia I isolectins. Biochemistry 18:4999-5005.
19. Lamb, J. E., I. J. Goldstein, F. L. Bookstein and L. E. Newton. 1980. Bandeiraea simplicifolia I isolectins reveal a precursor-product relationship, Plant Physiol. Suppl. 65:52 #280.
20. Ross, T. T., C. E. Hayes and I. J. Goldstein. 1976. Carbohydrate-binding properties of an immobilized α-D-galactopyranosyl-binding protein (lectin) from the seeds of Bandeiraea simplicifolia. Carbohydr. Res. 47:91-97.
21. Eckhardt, A. E. and I. J. Goldstein. 1979. An α-D-galactopyranosyl-containing glycoprotein from Ehrlich ascites tumor cell plasma membranes. Glycoconjugate Res. Vol. II. 1043-1045.
22. Peters, B. P. and I. J. Goldstein. 1979. The use of fluorescein-conjugated Bandeiraea simplicifolia B₄-isolectin as a histochemical reagent for the detection of α-D-galactopyranosyl groups, Exp. Cell Res. 120:321-334.
23. Stanley, W. S., B. P. Peters, D. A. Blake, D. Yep, E. H. Y. Chu and I. J. Goldstein. 1979. Interaction of wild-type and variant 3T3 cells with lectins from Bandeiraea simplicifolia seeds. Proc. Natl. Acad. Sci. USA. 76:304-307.

Chapter Three

STRUCTURE AND FUNCTION OF COMPLEX CARBOHYDRATES ACTIVE IN
REGULATING THE INTERACTIONS OF PLANTS AND THEIR PESTS

PETER ALBERSHEIM, MICHAEL McNEIL, ALAN G. DARVILL,
BARBARA S. VALENT, MICHAEL G. HAHN, BORRE K.
ROBERTSEN, AND PER ÅMAN

Department of Chemistry
University of Colorado
Boulder, Colorado 80309

INTRODUCTION

Our laboratory has recently come to realize that complex carbohydrates of higher plants, fungi, and bacteria can
act as regulatory molecules, that is, as molecules which in

minute quantities alter the metabolism of receptive cells by causing the synthesis of specific proteins. It is not surprising that these structurally complex and exquisitely specific molecules can possess regulatory properties, as many diverse classes of molecules including glycoproteins, proteins, peptides, steroids and a variety of smaller molecules such as epinephrine, indoleacetic acid, gibberellic acid, cytokinins, and even ethylene, are known to possess regulatory properties.

The carbohydrate portions of glycoconjugates are critically involved in recognition phenomena in biology (see, for example, references[1-3]). One of the earliest known functions of such carbohydrates in recognition processes is defining the blood group substances of mammals. The carbohydrate portions of glycoconjugates act on the cell surface of bacteria as serological determinants and as receptors for phage and bacteriocins. More recently, the carbohydrate portions of glycoconjugates have been recognized as cell surface specific antigens of at least some fungi, and the receptors of hormones and toxins in eucaryotic cells. The carbohydrate portions of glycoconjugates participate critically in determining the movement of glycoproteins within and between cells by acting as signals for directed transport of these molecules. Cell surface glycoconjugates are also important in differentiation of invertebrates, and they are the receptors for mitogenic lectins. Thus, the complex carbohydrates of glycoconjugates participate in a wide range of critical recognition phenomena. Evidence has even been obtained that the carbohydrate portion of a low molecular weight glycoconjugate has the ability to induce cell differentiation in a slime mold,[4] a function related to those which will be described in this paper.

In spite of knowing this wide range of recognition functions of the carbohydrate portion of glycoconjugates, there has been no suggestion until recently that carbohydrates themselves can act as regulatory molecules that alter protein synthesis in receptive cells, but recent evidence, accumulated with several experimental systems, establishes that carbohydrates can do just that. We describe below several plant and plant-microbe systems in which carbohydrates act as regulatory molecules.

PLANTS, WHEN EXPOSED TO CERTAIN β-GLUCAN FRAGMENTS OF FUNGAL
ORIGIN, DEFEND THEMSELVES BY SYNTHESIZING AND ACCUMULATING
PHYTOALEXINS

Many plants respond to invasion by a pathogenic or a
nonpathogenic microorganism, whether a fungus, a bacterium,
or a virus, by accumulating phytoalexins. Phytoalexins are
defined as low molecular weight antimicrobial compounds that
are both synthesized by and accumulate in plants after exposure to microorganisms. Many plants attempt to defend themselves against microbes and, perhaps, against other pests[5-13]
by producing phytoalexins. Molecules which trigger phytoalexin production in plants have been called elicitors.[14]

The best characterized and most effective known elicitor
of biological origin is composed of fragments of β-glucans
present in the mycelial walls of many fungi.[15] [16] This
"β-glucan elicitor" can be obtained by partial acid hydrolysis of purified mycelial walls of the fungal pathogen of
soybeans, Phytophthora megasperma f.sp. glycinea which causes
root stem rot. The "β-glucan elicitor" is very active in
stimulating phytoalexin accumulation in soybean tissues.
The smallest β-glucan fragments which have elicitor activity
contain approximately nine β-glucosyl residues interconnected
by 3-, 6-, and 3,6-glucosidic linkages.

The "β-glucan elicitors" isolated from different races
of Phytophthora megasperma f.sp. glycinea[5] and from the yeast
Saccharomyces cerevisiae[17] do not differ significantly in
their elicitation of phytoalexin accumulation in several
soybean cultivars,[5,17] in French beans,[18] and in potatoes.[18]
Thus, as is common for regulatory molecules, elicitors are
not species specific with regard to their source nor with
regard to the cells whose metabolism they regulate. Also
like other regulatory molecules, the "β-glucan elicitor" is
effective in very small concentrations; approximately ten
nanograms of the "β-glucan elicitor" stimulates accumulation
in a single soybean cotyledon of more than sufficient
amounts of phytoalexins to stop the growth of a variety of
microorganisms in vitro.

Evidence has been obtained that the "β-glucan elicitor"
like many other regulatory molecules, stimulates de novo
enzyme synthesis in receptive plant cells. The "β-glucan
elicitor" causes soybean cells to accumulate at least five

chemically and metabolically related pterocarpan phytoalexins.
Ebel, Hahlbrock, and Grisebach and their coworkers[19-21] have
studied the biosynthesis of these soybean phytoalexins.
Apparently, synthesis of these phenylpropanoid compounds is
a result of de novo synthesis of the necessary enzymes.
Dixon and his coworkers have shown that enzymes responsible
for the biosynthesis of the phytoalexin phaseollin in French
beans (Phaseolus vulgaris) are also synthesized de novo as
a result of elicitation by the "β-glucan elicitor".[18, 22, 23]
Thus, the "β-glucan elicitor" causes receptive cells to
synthesize new proteins.

POLYSACCHARIDE FRAGMENTS FROM THE WALLS OF PLANT CELLS ELICIT PHYTOALEXIN ACCUMULATION IN PLANT CELLS

Realization that the polysaccharides of the walls of
growing plant cells are extremely complex structures[24-26]
has made us wonder about the function of these molecules.
Until recently, these complex molecules had been thought to
have only a structural function, but it is difficult to
believe that such complex molecules have evolved for only
structural requirements. This skepticism proved well
founded for we have recently demonstrated that at least two
plant cell wall polysaccharides, or fragments thereof, serve
as regulatory molecules.

We have shown that one of these cell-wall derived
regulatory molecules, which elicits phytoalexins in soybean
cotyledons, is a component of isolated cell walls of soybean
stems and of the walls of suspension-cultured cells of
tobacco, sycamore, and wheat. This elicitor can be released
from the isolated walls by partial acid hydrolysis, and purified by ion exchange and gel filtration chromatography. The
elicitor-active fragments thus obtained are heterogeneous in
size. Their elution volume by gel chromatography suggests
that many of the elicitor-active fragments consist of 10 to
15 glycosyl residues. These elicitor-active fragments are
called the "endogenous elicitor".

The "endogenous elicitor" originates from a galacturonic
acid-rich cell wall polysaccharide; treatment of the "endogenous elicitor" with a endopolygalacturonase destroys its
elicitor activity. The "endogenous elicitor" of soybean
cell walls does not appear to originate from either rhamnogalacturonan I or rhamnogalacturonan II, the two pectic

polysaccharides which have been partially characterized in this laboratory.[24,26] This is not surprising as more than half of the pectic polysaccharides of the walls of growing cells have yet to be characterized.

We were not the first to discover that plants have an "endogenous elicitor". Bailey, Hargreaves and Selby[27-29] found a heat stable, dialyzable component which is released from damaged pea or bean tissues and which elicits phytoalexin accumulation in these tissues. It is not known whether the "constitutive" elicitor discovered by Bailey, Hargreaves and Selby is the same as the "endogenous elicitor" present in cell walls, but it seems likely that both elicitors are the same.

The realization that the "endogenous elicitor" is a fragment of a cell wall pectic polysaccharide is made more intriguing by observations that two enzymes which degrade pectic polysaccharides are elicitors. Stekoll and West[30] have studied an elicitor of casbene, a castor bean phytoalexin. The elicitor, present in culture filtrates of the pathogenic fungus Rhizopus stolonifer, is a pectic-degrading enzyme, an endopolygalacturonase. More recently, G. Lyon and P. Albersheim (unpublished results) obtained evidence that an elicitor secreted by the bacterial pathogen, Erwinia carotovora which causes soft rot, is a polygalacturonic acid lyase, another pectic-degrading enzyme. Partially purified preparations of this enzyme are effective elicitors of phytoalexin accumulation in soybean cotyledons.

The ability of pectic-degrading enzymes secreted by Rhizopus stolonifer and Erwinia carotovora to stimulate phytoalexin accumulation suggests that these enzymes could release the "endogenous elicitor" present in the cell walls of plants. However, we have not successfully released the "endogenous elicitor" of soybean cell walls by treatment of the walls with the E. carotovora polygalacturonic acid lyase. An alternative mechanism by which the pectic-degrading enzymes may elicit phytoalexin accumulation is indirectly by damaging plant cells.[31-34] The damaged cells might release or activate a plant enzyme which liberates the "endogenous elicitor". We have experimental support for this alternative explanation (G. Lyon and P. Albersheim, unpublished results). We have solubilized and partially purified an enzyme from soybean stems that elicits phytoalexin accumulation in soybean

cotyledons. This enzyme has only been isolated from stems
whose cells had been damaged by a freeze-thaw procedure.
Experiments have not yet been carried out to determine
whether this enzyme works by releasing the "endogeneous
elicitor". The activation of an elicitor-releasing enzyme in
damaged cells could explain the manner by which phytoalexin
accumulation is stimulated by a variety of abiotic elicitors
such as ultraviolet light, freeze-thawing, heavy metals, and
antibiotics, and perhaps even by the "β-glucan elicitor" .

The "endogenous elicitor" is likely to be distributed
throughout the plant. The enzyme, putatively responsible
for releasing the "endogenous elicitor", must be regulated
in some manner. The enzyme might be compartmentalized, such
as in lysosomes, or bound to the cell membrane, or stored in
an inactive form, perhaps as a zymogen. If this putative
enzyme is released or activated by cell damage and if this
enzyme is also distributed throughout the plant, all parts
of the plant would, as observed, be capable of localized
phytoalexin accumulation in response to any stimulus which
causes cell damage.

THE PROTEINASE INHIBITOR INDUCING FACTOR (PIIF) IS A FRAGMENT OF A CELL WALL POLYSACCHARIDE

A third complex carbohydrate found to be a regulatory
molecule is the plant hormone known as "PIIF" - the proteinase inhibitor inducing factor - which, like the "endogenous
elicitor", is a fragment of a polysaccharide present in the
walls of growing plant cells. Ryan and his coworkers[35]
discovered 10 years ago that the leaves of the potato and
tomato plants that had been attacked by the Colorado potato
beetle rapidly accumulate two proteinase inhibitors. The
proteinase inhibitors accumulate even in unattacked leaves
distant from the site of attack. The proteinase inhibitors
are proteins which have been purified to homogeneity and
well-characterized.[35]

Ryan and his coworkers found that insects are not
necessary for stimulation of inhibitors. Virtually any type
of extensive crushing or tearing of the vegetative tissues
of tomato, potato, and other dicotyledonous plants releases
PIIF into the vascular system of the plant where it is
transported to other tissues of the plant and initiates
accumulation of proteinase inhibitors.[35]

Ryan and his colleagues found that PIIF was heat stable, but they were unable to purify PIIF to homogeneity. Nevertheless, the properties of their partially purified preparations suggested that PIIF might be a carbohydrate. Our laboratory formed a collaboration with Ryan's group and analyzed their PIIF-active fractions for carbohydrate constituents.

The first of Ryan's preparations of PIIF-active material that our laboratory examined was impure and contained a variety of different glycosyl residues connected by a still larger variety of glycosyl linkages. However, this mixture contained those characteristically-linked glycosyl residues present in rhamnogalacturonan I,[26] a pectic polysaccharide accounting for approximately 7% of the walls of suspension-cultured sycamore cells. Assay of several different highly purified plant cell wall components for PIIF activity showed that rhamnogalacturonan I was the only component tested in tomatoes which possessed PIIF activity. Studies of more purified preparations of PIIF-active material extracted from tomato plants, and of other rhamnogalacturonan I preparations from sycamore have demonstrated that PIIF is, in fact, a fragment of rhamnogalacturonan I. Thus, just as with the "endogenous elicitor", it is evident that damage of plant cells releases fragments of a cell wall polysaccharide, in this case rhamnogalacturonan I or fragments thereof, which induces the synthesis in plant cells of proteins involved in defense of the plant (C. A. Ryan, P. Bishop, G. Pearce, A. Darvill, M. McNeil, and P. Albersheim, manuscript submitted).

PIIF-active rhamnogalacturonan I can be released from isolated cell walls by the action of a highly purified fungal endopolygalacturonase. The PIIF-active rhamnogalacturonan I has been purified by ion exchange and gel filtration chromatography. Purified rhamnogalacturonan I has a molecular weight of approximately 200,000 and is composed of $\underline{L}$-rhamnosyl, $\underline{D}$-galacturonosyl, $\underline{L}$-arabinosyl, and $\underline{D}$-galactosyl residues in the ratio of approximately 2:5:3:3. The backbone of rhamnogalacturonan is composed predominantly, if not entirely, of $\underline{D}$-galacturonosyl and $\underline{L}$-rhamnosyl residues. There are about 500 glycosyl residues in the backbone, but it is not known whether the backbone is a single contiguous glycan or whether each molecule contains a number of interconnected backbone chains. About half of the rhamnosyl residues of rhamnogalacturonan I are 2-linked, have a galacturonosyl residue attached to C-2, and are glycosidically

attached to C-4 of a galacturonosyl residue. The other half
of the rhamnosyl residues are 2,4-linked, have a galacturonosyl residue glycosidically attached at C-2, and are glycosidically attached to C-4 of a galacturonosyl residue. Side
chains averaging six glycosyl residues in length are attached
to C-4 of the 2,4-linked rhamnosyl residues. There are many
different side chains containing variously linked $\underline{L}$-arabinosyl
and/or $\underline{D}$-galactosyl residues. The size or even the composition of the smallest rhamnogalacturonan I fragment which
possesses PIIF activity is not known.

GLYCOPROTEINS SECRETED BY INCOMPATIBLE RACES (RACES THAT CAN
NOT INFECT THE PLANT) OF A FUNGAL PATHOGEN OF SOYBEANS ACT
AS REGULATORY MOLECULES AND PROTECT THE PLANT FROM ATTACK BY
COMPATIBLE RACES

     Almost all the microorganisms and other pests with which
a plant comes in contact cannot successfully pathogenize the
plant. The few microorganisms which are plant pathogens are
often highly specialized and are pathogenic on only one or a
few species of plants. Most "host-specific" pathogen species
have a number of races, each of which is distinct from the
others in its ability to attack various varieties (cultivars)
of its host plant species. In other words, race 1 of a
pathogen of a particular crop may attack variety A but not
variety B, while race 2 of the pathogen may attack variety B
but not variety A. Both races might be able to attack
variety C and neither variety D, and so on. In this type of
host-pathogen system, for each gene that governs resistance
in the host plant there is a corresponding gene in the fungal
pathogen that governs avirulence. This type of relationship
is referred to in the plant pathology literature as a gene-
for-gene host-pathogen system.[36-38]

     Gene-for-gene resistance in plants is determined by
dominant Mendelian genes.[36-39] Each such resistance gene
that a plant possesses can make the plant totally resistant
to one or more races of at least one of its pathogens. However, a resistance gene is effective in protecting a plant
against only those pathogen races which produce molecules
capable of a specific interaction with the product of the
resistance gene. Since these molecules of the pathogen
cause the pathogen to be avirulent, the genes responsible
for the synthesis of these molecules are called avirulence
genes rather than virulence genes.

The interdependence of resistance and avirulence genes leads to the conclusion that the products of specific resistance genes of the host must recognize (interact with) the products of specific avirulence genes of the pathogen. This recognition reaction is the key to whether a race of gene-for-gene pathogens will be compatible with (virulent on) a variety of its host.[1,36] A positive interaction of a product of a resistance gene with the product of an avirulence gene initiates a resistance or incompatible response in the plant.

We have hypothesized that the avirulence genes of a gene-for-gene pathogen are manifest as cell surface or extracellular structures.[1] The only fungi whose surface structures have been extensively studied are the yeasts. Ballou et al.[40,41] have demonstrated that in yeast the immunodominant species-specific cell surface structures are portions of mannan-containing glycoproteins. The species-specific differences in the glycoproteins reside in small differences in the structures of the carbohydrate portion of these glycoproteins.

At least some of the enzymes secreted by yeast are themselves mannan-containing glycoproteins, and the structures of the mannan portions include the same antigenically active carbohydrate structures as the species-specific cell surface mannan-containing glycoproteins.[42,43] The carbohydrate portions of the cell surface and extracellular glycoproteins are synthesized by the same glycosyl transferases.[40] Thus, each species of yeast has a unique set of glycosyl transferases that is responsible for the synthesis of these species-specific antigens.

We have suggested that the products of the avirulence genes of fungal pathogens are glycosyl transferases, enzymes which function in the synthesis of complex carbohydrates which are present both on fungal cell surface and on at least some secreted glycoproteins. We think of the products of a plant's resistance genes as receptors for the glycoproteins synthesized by the avirulence gene-encoded glycosyl transferases of the pathogen. Thus, we propose that complex carbohydrates, present on the cell surface and/or extracellular glycoproteins of the pathogen, are recognized by receptors in resistant varieties of the pathogen's host and that this interaction activates the host's defenses. If the

hypothesis is correct and if the plant pathogenic fungi are
similar in this respect to yeast, at least some of the glycoproteins secreted by a pathogen will contain race-specific
complex carbohydrates.

We have been studying the race- and cultivar-specific
interaction of soybeans and Phytophthora megasperma f.sp.
glycinea, the causal agent of root and stem rot. This host-
pathogen system appears to be a gene-for-gene system, since
there exist at least 16 fungal races and many differently
susceptible cultivars of the host plant.[44] Invertase, which
is one of the many proteins secreted by this pathogen, was
chosen for study as a typical extracellular protein of this
pathogen. As with yeast, the invertases secreted by races
1, 2, and 3 of Phytophthora megasperma f.sp. glycinea are
mannan-containing glycoproteins.[45] The glycosyl linkage
compositions of the carbohydrate portions of the invertases
produced by three different Phytophthora races are clearly
different.[45] The demonstration of race-specific carbohydrate structures in differentially virulent Phytophthora
races provides support for the hypothesis that such complex
carbohydrates are involved in determining specificity in
gene-for-gene host-pathogen systems, for the only known way
to discriminate between the races and the only known selection pressure to cause differences in the races is by their
differing abilities to infect the various soybean cultivars.

We reasoned that if the race-specific carbohydrates of
the extracellular glycoprotein population determine host-
pathogen specificity, the biological activity of these molecules should be demonstrable. In other words, the extracellular glycoproteins from incompatible (avirulent) races of
Phytophthora megasperma f.sp. glycinea, but not those from
compatible (virulent) races, should be capable of activa-
ting a defense reaction in seedlings of a soybean cultivar
which would thereby protect the seedlings from attack by
compatible races of the fungal pathogen.

Our approach to demonstrating the biological activity
of the extracellular glycoproteins was first to partially
purify the glycoprotein fraction from the extracellular
culture medium of three races of Phytophthora megasperma
f.sp. glycinea. The macromolecules obtained were composed
on the average of 81.5% protein and 18.5% carbohydrate.
Analysis of the carbohydrate fractions showed quantitative

but not qualitative differences in their composition.[46] This result is similar to that obtained for the carbohydrate fractions of the extracellular invertases of these three races.[45]

The important question was whether the extracellular glycoproteins of incompatible Phytophthora races can protect soybean cultivars from compatible races of this fungal pathogen. This would be expected if these glycoproteins are the race-specific determinants, that is, the biochemical expression of the avirulence genes. Experiments to answer this important question were carried out with three races of Phytophthora megasperma f.sp. glycinea and four soybean cultivars that are differentially susceptible or resistant to the races of Phytophthora. In the combinations tested, the extracellular glycoproteins from incompatible, but not from compatible, races of Phytophthora megasperma f.sp. glycinea protect seedlings from infection by compatible races of the pathogen. For example, the extracellular glycoproteins from races 1 or 2 protect the cultivar Harosoy 63, with which races 1 and 2 are incompatible, from infection by race 3, while the extracellular glycoproteins from compatible race 3 do not protect the Harosoy 63 seedlings from race 3 fungus.[46] On the other hand, the extracellular glycoproteins from races 1 or 3 protect the cultivar Sanga, with which races 1 and 3 are incompatible, from the compatible race 2 fungus, although the extracellular glycoproteins from race 2, which do protect Harosoy 63 from race 3, do not protect Sanga from race 2.[46]

These positive results encouraged us. We are even more encouraged by results of our first protection experiments with the alkali-released carbohydrate portion of the race-specific glycoproteins. These experiments tentatively indicate that the carbohydrate portions, by themselves, are more effective race-specific protection factors than the intact glycoproteins.

A long term goal is to look for receptors in the soybean seedlings for the specificity factors. If our hypothesis is correct, the receptors should be present in those plants which interact with the race-specific glycoproteins, that is, in resistant cultivars, but should not be present in plants which do not interact with the race-specific glycoproteins, that is, in susceptible cultivars. The receptors are likely

to be the products of the resistance genes of the soybean
cultivars.

ACIDIC POLYSACCHARIDES SECRETED BY THE SYMBIOTIC NITROGEN-
FIXING RHIZOBIA APPEAR TO REGULATE THE ENTRY OF THESE BAC-
TERIA INTO THE ROOTS OF LEGUMES

     A great many publications have demonstrated the essen-
tial function of the cell surface and extracellular polysaccharides of Gram-negative bacteria in the interaction of
these bacteria with other cells, including the cells of both
plants and animals.[1,47,48] The nitrogen-fixing Rhizobium
are Gram-negative bacteria, therefore their surface and
extracellular polysaccharides are likely to be active as
regulatory molecules by determining with which higher plants
the Rhizobium can form symbiotic nitrogen-fixing relationships. This hypothesis has been supported by the results
described in this section.

     Numerous extracellular polysaccharides of Gram-negative
bacteria have been structurally characterized, and these polysaccharides are, in general, serotype or species specific.
This also appears to be generally true for Rhizobium species
for, with the possible exception of R. leguminosarum and R.
trifolii, the acidic polysaccharides secreted by the various
Rhizobium species appear to be nodulation group specific,
that is, Rhizobium which nodulate different legumes secrete
different extracellular polysaccharides. For example, R.
japonicum[49] which nodulates and fixes nitrogen in soybeans
secretes a markedly different acidic polysaccharide than
does R. meliloti[50] which nodulates alfalfa. R. leguminosarum,
the pea symbiont, R. trifolii, the clover symbiont, and R.
phaseoli, the true bean symbiont, are the most closely related Rhizobium species.[51,52] We have found that R. phaseoli
secretes an acidic polysaccharide with a different structure
than that secreted by R. leguminosarum and R. trifolii (P.
Åman, L.-E. Franzén, M. McNeil, A. Darvill and P. Albersheim,
unpublished results). However, we have also shown that the
acidic polysaccharides secreted by R. leguminosarum and R.
trifolii have basically the same structures.[53] The structure of the acidic extracellular polysaccharides produced by
R. leguminosarum and R. trifolii, both fast-growing rhizobia,
has some similarities to the acidic polysaccharides secreted
by R. meliloti and R. phaseoli, which are also fast-growing
species. However, there is no relationship between the

structures of the acidic polysaccharides secreted by the
fast-growing Rhizobium species and the acidic polysaccharide
secreted by slow-growing R. japonicum.

We have established that the glycosyl residue sequence
and the anomeric configurations of the glycosyl linkages of
the acidic polysaccharides secreted by two R. trifolii and
two R. leguminosarum strains are identical. We have not,
however, investigated the possibility of differently substituted acetyl, succinyl, or other alkali-labile residues in
these polysaccharides. Therefore, it is not yet established
that the acidic extracellular polysaccharides from R.
leguminosarum and R. trifolii have identical structures.
Analyses for alkali-labile substituents are important, for
Jansson et al.[50] have determined that the acidic polysaccharide secreted by R. trifolii U226 possesses at least one
0-acetyl residue per repeating unit; the 0-acetyl residue(s)
is attached to C-2 and/or C-3 of a 4-linked glucosyl residue(s).
It remains to be ascertained whether the R. leguminosarum
polysaccharide has the same acetyl substitution.

R. leguminosarum is a symbiont of pea (Pisum sativum)
and R. trifolii a symbiont of clover (Trifolium pratense).
In some instances these two species cross nodulate their
legume hosts. Some strains of R. leguminosarum nodulate
Trifolium species and some strains of R. trifolii nodulate
Pisum species,[54-57] although the nodules formed in each case
are unable to fix nitrogen. Both R. leguminosarum and R.
trifolii cause curling and branching of root hairs in a host
of R. trifolii, Trifolium glomeratum, phenomena generally
induced only by Rhizobium capable of forming a symbiosis
with that legume.[57, 58] R. leguminosarum and R. trifolii have
also been reported to have a high degree of homology between
their DNA molecules.[51] Thus, it is not very surprising that
the polysaccharides secreted by R. leguminosarum and R.
trifolii are very similar if not identical. The facts that
R. leguminosarum and R. trifolii can in some instances cross
nodulate legumes and that these two Rhizobium species secrete
identical or nearly identical acidic polysaccharides support
the hypothesis that the secreted acidic polysaccharides
participate as regulatory molecules in the recognition processes which permit rhizobia to enter their host legumes.

Strong support for a regulatory function of acidic polysaccharides secreted by Rhizobium species is provided by

findings of W. D. Bauer and his coworkers at the Charles
Kettering Institute. They have obtained evidence that the
acidic polysaccharides are required for the development of
legume root hairs capable of being infected by symbiont
rhizobia; root hairs developed in the absence of these
polysaccharides can not be infected (ref. 59 and personal
communication). Therefore, these polysaccharides constitute
a good example of complex carbohydrates with regulatory properties, in this case an ability to cause a specific differentiation of the epidermal cells of legume roots.

RHAMNOGALACTURONAN II - AN EXTRAORDINARILY COMPLEX POLYSAC-
CHARIDE PRESENT IN THE WALLS OF GROWING PLANT CELLS

We have recently isolated, from the walls of suspension-
cultured sycamore cells, a previously unknown pectic polysaccharide called rhamnogalacturonan II.[24] Rhamnogalacturonan II, which accounts for about 4% of the cell wall, is
completely solubilized from the walls of suspension-cultured
sycamore cells by the action of an endo-α-1,4-polygalact-
uronase and separated from the other solubilized pectic
polysaccharides by anion exchange and gel permeation chromatography.

A brief description of rhamnogalacturonan II is
included here because the extreme structural complexity of
this molecule suggests that it too will be found to function
as a regulatory molecule. This polysaccharide, which has
been purified to apparent homogeneity, possesses a well-
defined structure and molecular size. As isolated, rhamnogalacturonan II contains a total of about 50 glycosyl residues. It contains nine different glycosyl constituents
including the rarely observed sugars 2-O-methyl fucose,
2-O-methyl xylose, which have nevertheless previously been
recognized to be trace components of pectic polymers,[60,63]
and apiose, a branched pentose, which has also been recognized as a component of the pectic polysaccharides of Lemna
species. The Lemna-type apiose-containing pectic polysaccharide has not been found to be widespread in nature and is
not structurally related to rhamnogalacturonan II. Apiose
and the 2-O-methyl derivatives of fucose and xylose have
never previously been recognized to be associated in a single
polysaccharide, although all three sugars have been isolated
from leaves of deciduous trees.[64]

Rhamnogalacturonan II is characterized by many different terminal glycosyl residues including terminal galacturonosyl, terminal galactosyl, terminal arabinosyl, terminal 2-0-methyl xylosyl, terminal 2-0-methyl fucosyl, and terminal rhamnosyl residues. The large content of terminal glycosyl residues and of a variety of branched glycosyl residues indicates a highly branched structure. Rhamnogalacturonan II also contains a number of unusually linked glycosyl residues including 2-linked glucuronosyl, 3´-linked apiosyl, 3-linked rhamnosyl, 2,4-linked galactosyl, and 3,4-linked fucosyl residues. The glycosyl composition of rhamnogalacturonan II remains constant throughout the lag, log, and stationary phases of growth of suspension-cultured sycamore cells. We also have evidence that a molecule very similar or identical to rhamnogalacturonan II is present in the primary cell walls of the four other dicots examined; namely, pea, French bean, and tomato seedlings, and suspension-cultured tobacco cells.

It is interesting to consider how a polysaccharide as complex as rhamnogalacturonan II is synthesized; synthesis by any of the known pathways would require on the order of 100 enzymes. This is an enormous investment by the cell to achieve structural complexity in a polymer that represents only 4% of the wall. Why is there such an investment? We can't help but think that the reason has been to evolve a molecule with regulatory functions. We are very curious to learn the function of this molecule.

CONCLUDING REMARKS

It has been an exciting experience for us to realize that the plant cell wall polysaccharides, whose structures we have been struggling to decipher, are functioning not only as structural polymers but also in a regulatory capacity. At least two of the complex polysaccharides which are present in the walls of growing plant cells contain fragments which possess the remarkable properties of hormones, that is, molecules formed by one cell which in minute amounts, stimulate receptive cells to synthesize specific proteins. Two different pectic polysaccharides contain within themselves the glycosyl sequences which constitute either the plant hormone known as PIIF or the "endogenous elicitor". PIIF and the "endogenous elicitor" are apparently released from the cell walls surrounding injured cells and then stimulate receptive cells to synthesize proteins involved in defense of the plant.

Two of the other complex carbohydrates described in this paper, the "β-glucan elicitor" and the acidic polysaccharides secreted by Rhizobium species, also appear to possess the attributes of hormones, except that these regulatory carbohydrates are produced by one organism and affect receptive cells in another organism.

PIIF, the "endogenous elicitor" , and the "β- glucan elicitor" have, in addition to being carbohydrates with regulatory properties, two other characteristics in common. They are "stored" as insoluble cell wall polysaccharides; and they are released or "activated" by cleavage of the wall polysaccharides, presumably by specific enzymes.

The fact that the walls of growing plant cells contain these oligosaccharide "hormones" means that the walls function as a "pseudogland" containing regulatory molecules which can be released as needed. We can envision the cell wall containing a variey of messages capable of controlling physiological processes of a developing plant.

The fact that these oligosaccharide hormones originate as portions of larger polymers is strikingly similar to the origin of a number of animal peptide hormones. For example, several polypeptide hormones synthesized in the pituitary gland originate in a common precursor polypeptide.[65] A 16,000 dalton fragment of the precursor polypeptide is removed and the remaining polypeptide cleaved to produce the hormones corticotropin and β-lipotropin. The corticotropin can be further cleaved to yield α-melanotropin, and the β-lipotropin can be cleaved to yield α-lipotropin and β-endorphin. This process is analogous to cleavage of a wall polysaccharide to yield biologically active fragments.

We hope that the knowledge that plant cell wall polysaccharides possess a number of interesting biological functions will stimulate other laboratories to focus on unraveling their complex structures. Certainly, the increasing realization that complex carbohydrates play key roles in biological recognition processes will stimulate efforts to develop rapid and efficient methods for the structural characterization and synthesis of these molecules.

ACKNOWLEDGEMENTS

The excellent professional technical assistance of David J. Gollin, Constance L. Winans, and William S. York is gratefully acknowledged.

The research reported here has been supported by grants from the U. S. Dept. of Energy (EY-76-S-02-1426), The Rockefeller Foundation (GA COH 7904 and RF78035), the U. S. Dept. of Agriculture (5901-0410-8-0194-0 and COL-616-15-73) and the National Science Foundation (PCM79-94491).

REFERENCES

1. Albersheim, P. and A. J. Anderson-Prouty. 1975. Carbohydrates, proteins, cell surfaces, and the biochemistry of pathogenesis. Annu. Rev. Plant Physiol. 26:31-52.
2. Lerner, R. A. and D. Bergsma. 1978. The Molecular Basis of Cell-Cell Interaction, XIV (2). Liss, New York.
3. Marchesi, V. T., V. Ginsburg, P. W. Robbins, and C. F. Fox. 1978. Cell Surface Carbohydrates and Biological Recognition. Liss, New York.
4. Town, C. and E. Stanford. 1979. An oligosaccharide-containing factor that induces cell differentiaton in Dictyostelium discoideum. Proc. Natl. Acad. Sci. USA. 76:308-312.
5. Albersheim, P. and B. S. Valent. 1978. Host-pathogen interactions in plants. Plants, when exposed to oligosaccharides of fungal origin, defend themselves by accumulating antibiotics. J. Cell Biol. 78:627-643.
6. Ayers, A. R., J. Ebel, B. Valent, and P. Albersheim. 1976. Host-pathogen interactions X. Fractionation and biological activity of an elicitor isolated from the mycelial walls of Phytophthora megasperma var. sojae. Plant Physiol. 57:760-765.
7. Deverall, B. 1977. Defense Mechanisms of Plants. Cambridge University Press, London.
8. Grisebach, H. and J. Ebel. 1978. Phytoalexins, chemical defense substances of higher plants? Angew. Chem. Int. Ed. Engl. 17:635-647.
9. Ingham, J. L. 1972. Phytoalexins and other natural products as factors in plant disease resistance. Bot. Rev. 38:343-424.

10. Keen, N. and B. Bruegger. 1977. Phytoalexins and
     chemicals that elicit their production in plants. In:
     Host Plant Resistance to Pests. (P. A. Hedin, ed.).
     ACS Symp. Ser. 62, Washington, D. C. pp. 1-26.
11. Kuć, J. 1972. Phytoalexins. Annu. Rev. Phytopathol.
     10:207-232.
12. Kuć, J. and W. Currier. 1976. Phytoalexins, plants,
     and human health. Adv. Chem. Ser. 149:356-368.
13. VanEtten, H. and S. Pueppke. 1976. In: Biochemical
     Aspects of Plant-Parasitic Relationships. J. Friend
     and D. Threlfall, eds. Academic Press, London.
     pp. 239-389.
14. Keen, N. T., J. E. Partridge, and A. I. Zaki. 1972.
     Pathogen-produced elicitor of a chemical defense
     mechanism in soybeans monogenically resistant to
     Phytophthora megasperma var. sojae. Phytopathology
     62:768.
15. Bartnicki-Garcia, S. 1968. Cell wall chemistry,
     morphogenesis, and taxonomy of fungi. Annu. Rev.
     Microbiol. 22:87-108.
16. Seitsma, J. H. and J. G. H. Wessels. 1977. Chemical
     analysis of the hyphal wall of Schizophyllum commune.
     Biochem. Biophys. Acta 496:225-239.
17. Hahn, M. and P. Albersheim. 1978. Host-pathogen
     interactions XIV. Isolation and partial characteri-
     zation of an elicitor from yeast extract. Plant Physiol.
     62:107-111.
18. Cline, K., M. Wade, and P. Albersheim. 1978. Host-
     pathogen interactions XV. Fungal glucans which elicit
     phytoalexin accumulation in soybean also elicit the
     accumulation of phytoalexins in other plants. Plant
     Physiol. 62:918-921.
19. Ebel, J., B. Schaller-Hekeler, K.-H. Knobloch, E.
     Wellman, H. Grisebach and K. Hahlbrock. 1974. Coor-
     dinated changes in enzyme activities of phenylpropanoid
     metabolism during the growth of soybean cell suspension
     cultures. Biochem. Biophys. Acta 362:417-424.
20. Zahringer, U., J. Ebel, and H. Grisebach. 1978. In-
     duction of phytoalexin synthesis in soybean. Elicitor-
     induced increase in enzyme activities of flavonoid
     biosynthesis and incorporation of mevalonate into
     glyceollin. Arch. Biochem. Biophys. 188:450-455.
21. Zähringer, U., J. Ebel, L. J. Mulheirn, R. L. Lyne,
     and H. Grisebach. 1979. Induction of phytoalexin
     synthesis in soybean. FEBS Lett. 101:90-92.

22. Dixon, R. A. and D. S. Bendall. 1978. Changes in the levels of enzymes of phenylpropanoid and flavonoid synthesis during phaseollin production in cell suspension cultures of Phaseolus vulgaris. Physiol. Plant Pathol. 13:295-306.

23. Dixon, R. A. and C. J. Lamb. 1979. Stimulation of de novo synthesis of L-phenylalanine ammonia-lyase in relation to phytoalexin accumulation in Colletotrichum lindemuthianum elicitor-treated cell suspension cultures of French bean (Phaseolus vulgaris). Biochem. Biophys. Acta 586:453-463.

24. Darvill, A., M. McNeil, and P. Albersheim. 1978. The structure of plant cell walls VIII. A new pectic polysaccharide. Plant Physiol. 62:418-422.

25. Darvill, J., M. McNeil, A. G. Darvill, and P. Albersheim. 1980. Structure of plant cell walls XI. Glucuronoarabinoxylan, a second hemicellulose in the primary cell walls of suspension-cultured sycamore cells. Plant Physiol. 66:1135-1139.

26. McNeil, M., A. G. Darvill, and P. Albersheim. 1980. Structure of plant cell walls X. Rhamnogalacturonan I, a structurally complex pectic polysaccharide in the walls of suspension-cultured sycamore cells. Plant Physiol. 66:1128-1134.

27. Bailey, J. A. 1970. Pisatin production by tissue cultures of Pisum sativum L. J. Gen. Microbiol. 61:409-415.

28. Hargreaves, J. A. and J. A. Bailey. 1978. Phytoalexin formation in cell suspensions of Phaseolus vulgaris in response to constitutive metabolites released by damaged bean cells. Physiol. Plant Pathol. 13:89-100.

29. Hargreaves, J. A. and C. Selby. 1978. Phytoalexin formation in cell suspensions of Phaseolus vulgaris in response to an extract of bean hypocotoyls. Phytochemistry 17:1099-1102.

30. Stekoll, M. and C. A. West. 1978. Purification and properties of an elicitor of castor bean phytoalexin from culture filtrates of the fungus Rhizopus stolonifer. Plant Physiol. 61:38-45.

31. Bateman, D. F. 1976. Plant cell wall hydrolysis by pathogens. In: Biochemical Aspects of Plant-Parasite Relationship. (J. Friend and D. R. Threlfall, eds.). Academic Press, London. pp. 79-103.

32. Bateman, D. F. and H. G. Basham. 1976. Degradation
    of plant cell walls and membranes by microbial enzymes.
    In: Encyclopedia of Plant Physiology, New Series, Vol.
    4, Physiological Plant Pathology. (R. Heitefuss and
    P. H. Williams, eds.). Springer-Verlag, Berlin.
    pp. 316-355.
33. Gardner, J. M. and C. I. Kado. 1976. Polygalacturonic
    acid trans-eliminase in the osmotic shock fluid of
    Erwinia rubrifaciens: Characterization of the purified
    enzyme and its effect on plant cells. J. Bacteriol.
    127:451-460.
34. Wood, R. K. S. 1977. Killing of protoplasts by plant
    pathogens. Proc. Symp. Hungar. Acad. Sci. pp. 107-115.
35. Ryan, C. A. 1978. Proteinase inhibitors in plant
    leaves: A biochemical model for pest-induced natural
    plant protection. Trends Biochem. Sci. July:148-150.
36. Day, P. R. 1974. Genetics of Host-Parasite Interaction.
    Freeman, San Francisco.
37. Flor, H. H. 1956. The complementary genic systems in
    flax and flax rust. Advan. Genet. 8:29-54.
38. Flor, H. H. 1971. Current status of the gene-for-
    gene concept. Annu. Rev. Phytopathol. 9:275-296.
39. Hooker, A. L. and K. M. S. Saxena. 1971. Genetics of
    disease resistance in plants. Annu. Rev. Genet.
    5:407-423.
40. Ballou, C. E. 1974. Some aspects of the structure,
    immunochemistry and genetic control of yeast mannans.
    In: Advances in Enzymology, Vol. 40. (A. Meister, ed.).
    Wiley, New York. pp. 239-270.
41. Ballou, C. E. and W. C. Raschke. 1974. Polymorphism
    of the somatic antigen of yeast. Science 184:127-134.
42. Biely, P., Z. Krátký and S. Bauer. 1976. Interaction
    of conconavalin A with external mannan-proteins of
    Saccharomyces cerevisiae. Eur. J. Biochem. 70:75-81.
43. Smith, W. L. and C. E. Ballou. 1974. Immunochemical
    characterization of the mannan component of the exter-
    nal invertase (β-fructofuranosidase) of Saccharomyces
    cerevisiae). Biochemistry 13:355-361.
44. Laviolette, F. A. and K. L. Athow. 1977. Three new
    physiologic races of Phytophthora megasperma var.sojae.
    Phytopathology 67:267-268.
45. Ziegler, E. and P. Albersheim. 1977. Host-pathogen
    interactions XIII. Extracellular invertases secreted
    by three races of a plant pathogen are glycoproteins
    which possess different carbohydrate structures. Plant
    Physiol. 59:1104-1110.

46. Wade, M. and P. Albersheim. 1979. Race-specific
     molecules that protect soybeans from Phytophthora
     megasperma var. sojae. Proc. Natl. Acad. Sci. USA.
     76:4433-4437.
47. Jann, K. and O. Wesphal. 1975. The Antigens.
     Academic Press, London.
48. Ørskov, I., F. Ørskov, B. Jann, and K. Jann. 1977.
     Serology, chemistry, and genetics of O and K antigens
     of Escherichia coli. Bactiol. Rev. 41:667-710.
49. Dudman, W. F. 1978. Structural studies of the extra-
     cellular polysaccharides of Rhizobium japonicum strains
     71A, CC708 and CB1795. Carbohydr. Res. 66:9-23.
50. Jansson, P.-E., L. Kenne, B. Lindberg, H. Ljunngren, J.
     Lönngren, U. Rudén, and S. Svensson. Demonstration of
     an octasaccharide repeating unit in the extracellular
     polysaccharide of Rhizobium meliloti by sequential
     degradation. J. Amer. Chem. Soc. 99:3812-3815.
51. Johnston, A. W. B. and J. E. Beringer. 1977.
     Chromosomal recombination between Rhizobium species.
     Nature 267:611-613.
52. Vincent, J. M. 1977. A Treatise on Dinitrogen Fix-
     ation. Wiley, New York. pp. 277-366.
53. Robertsen, B., P. Åman, A. G. Darvill, M. McNeil, and
     P. Albersheim. 1981. Host-symbiont interactions V.
     The Structure of the acidic extracellular polysac-
     charides secreted by Rhizobium leguminosarum and
     Rhizobium trifolii. Plant Physiol., in press.
54. Hepper, C. M. 1978. Physiological studies on nodule
     formation. The characteristics and inheritance of
     abnormal nodulation of Trifolium pratense by Rhizobium
     leguminosarum. Ann. Bot. 42:109-115.
55. Hepper, C. M. and L. Lee. 1979. Nodulation of
     Trifolium subterraneum by Rhizobium leguminosarum.
     Plant Soil 51:441-445.
56. Mleczhowska, J., P. S. Nutman, and G. Bond. 1944.
     Note on the ability of certain strains of Rhizobia from
     peas and clover to infect each other's host plants. J.
     Bacteriol. 48:673-675.
57. Vincent, J. M. The Biology of Nitrogen Fixation.
     Amer. Elsevier, New York. pp. 265-341.
58. Yao, P. Y. and J. M. Vincent. 1969. Host specificity
     in the root hair "curling factor" of Rhizobium spp.
     Aust. J. Biol. Sci. 22:413-423.

59. Bauer, W. D., T. V. Bhuvaneswari, A. J. Mort, and
    G. Turgeon. 1979. The initiation of infections in
    soybean by Rhizobium. 3. R. japonicum polysaccharide
    pretreatment induces root hair infectibility. Plant
    Physiol. (supplement) 63:135.
60. Aspinall, G. O. and A. Canas-Rodriquez. 1958. Sisal
    pectic acid. J. Chem. Soc. 4020.
61. Aspinall, G. O. and R. S. Fanshawe. 1961. Pectic
    substances from lucerne (Medicagi sativa) I. Pectic acid.
    J. Chem. Soc. (C), 4215.
62. Aspinall, G. O., K. Hunt and I. M. Morrison. 1966.
    Polysaccharides of soybeans. Part II. Fractionation
    of hull cell-wall polysaccharides and the structure of
    a xylan. J. Chem. Soc. (C), 1945-1949.
63. Barrett, A. J. and D. H. Northcote. 1965. Apple fruit
    pectic substances. Biochem. J. 94:617-627.
64. Bacon, J. S. D. and M. V. Cheshire. 1971. Apiose and
    mono-O-methyl sugars as minor constituents of the
    leaves of deciduous trees and various other species.
    Biochem. J. 124:555-562.
65. Liotta, A. S., C. Loudes, J. F. McKelvy, and D. T.
    Krieger. 1980. Biosynthesis of precursor corticotro-
    pin/endorphin-, corticotropin-, α-melanotropin-,
    β-lipotropin-, and β-endorphin-like material by cultured
    neonatal rat hypothalamic neurons. Proc. Natl. Acad. Sci.
    USA. 77:1880-1884.

Chapter Four

GALACTOFURANOSYL-CONTAINING LIPOGLYCOPEPTIDE IN PENICILLIUM

J. E. GANDER AND CYNTHIA J. LAYBOURN

Department of Biochemistry
College of Biological Sciences
University of Minnesota
St. Paul, Minnesota 55108

INTRODUCTION

Fungi secrete numerous macromolecules which are derived
from a) walls, b) enzymes, and c) subcellular organelles.
These macromolecules are subject to the action of lytic
enzymes located in the nutrient medium. Therefore, many polymers found in media supporting fungal growth are degradation
products of more complex substances. For instance, galactocarolose, a 5-O-β-$\underline{D}$-galactofuranosyl-containing decasac-
charide, and mannocarolose, an α-$\underline{D}$-mannopyranosyl-containing
nonasaccharide, first isolated from 28-day culture filtrates
of Penicillium charlesii and partially characterized in W. N.
Haworth's laboratory[1,2] have been shown to be derived from
a more complex glycopeptide[3-5] which first appears in the
growth medium soon after formation of conidia.[6] Because of
the composition of the complex glycopeptide we have referred
to it as a peptidophosphogalactomannan.[3] The glycopeptide
may be the major polysaccharide-containing substance secreted

prior to general lysis of the fungus.[6,7] Peptidophosphogalactomannans and/or peptidogalactomannans have been
obtained from Cladosporium werneckii,[8,9] species of Aspergillus,[10-12] several species of dermatophytes from the
genera of Trichophyton and Microsporum,[13,14] Fulvia fulva
(Cooke) Ciferri[15] and several species of Penicillium[10,16,17]
and may be common constituents of many genera of fungi.

It was initially assumed that polysaccharides found
in media supporting fungal growth were derived from cell
walls. This assumption may be correct for those polysaccharides which accumulate in large quantities in culture
media over a period of two to four weeks. However, the
peptidophosphogalactomannans and peptidogalactomannans
derived from two to four grams (dry weight) of mycelia constitute less than one percent of the total mass. These
glycopeptides are not major cell wall mannans or galactomannans. Cell walls of 14-day stationary cultures of P.
charlesii contain no galactofuranosyl residues[18] and no
phosphogalactomannan or galactomannan was released from cell
walls obtained from 3-day cultures of P. charlesii when
these walls were treated with alkali.[19] Bartnicki-Garcia[20]
divided 13 taxonomic groups of fungi and yeasts into 8 different categories on the basis of the composition of the
major polysaccharides of cell walls. Cell walls of Ascomycetes, the class to which Penicillium, Aspergillus,
Neurospora, Trichophyton, Alternaria, Fusarium and Hansenula
belong, contain an inner layer of chitin surrounded by an
outer layer of β-glucan.[21,22] Hunsley and Burnett[23] suggested that a layer of protein or glycoprotein resided between the layers of chitin and β-glucan. The composition of
the protein or glycoprotein has not been established.

Because the glycopeptides are somewhat unique in that
they contain 5-O-β-$\underline{D}$-galactofuranosyl residues and at least
those from P. charlesii contain 2-aminoethanol[4] and
2-dimethylaminoethanol[24] residues which are attached to the
mannan through phosphodiester linkage,[5] we initiated a
search for intracellular galactofuranosyl-containing polymers.
The objective is to determine a) the parent polymer from
which exocellular peptidophosphogalactomannan is derived,
and b) the location of the parent polymer and its function in
the fungus. This chapter reviews the work from this and
other laboratories resulting in the partial description of
the primary structure of this class of exocellular

glycopeptides, work from this laboratory on the location, isolation, and partial characterization of a membrane-bound galactofuranosyl-containing polymer, and a suggested function for the extracellular glycopeptide.

PARTIAL CHARACTERIZATION OF EXOCELLULAR FUNGAL GLYCOPEPTIDES

Investigations on structure and function of fungal glycopeptides originated in laboratories interested in the causative agent(s) of the immediate and delayed immunological responses observed when extracts of Trichophyton mentagrophytes or the polymers obtained from media supporting Trichophyton mentagrophytes growth were tested in patients suffering from tinea caused by Trichophyton species.[25,27] The immediate response of the glycopeptides, collectively named "trichophytin", was shown to be caused by the polysaccharide portion and the delayed response was due to the polypeptide(s).[27] Trichophytin obtained from submerged cultures contained a galactomannan, galactose:mannose ratio from 1:3 to 1:8, attached to a polypeptide which constituted about 10 percent of the total mass. In contrast, the polysaccharide of the glycopeptide obtained from surface cultures was shown to be a glucomannan.[25] Galactomannans contained numerous nonreducing terminal galactofuranosyl residues and the mannopyranosyl residues were attached predominantly by 1-2 and 1-4 linkages. They resolved trichophytin on DEAE-cellulose into two glycopeptides which had differing amino acid composition. Related galactomannans have been obtained from other species of Trichophyton and Microsporum.[14,28,29] Alkali effected the release of galactomannan and mannobiosyl residues from the polypeptide of Trichophyton granulosum.[14] The galactomannan contained 16 percent galactose which occurred as nonreducing terminal galactofuranosyl residues, and 84 percent mannopyranosyl residues which were attached by 1-2 and 1-6 linkages. Galactomannans were also obtained from Trichophyton schonlёnii, Trichophyton rubrum, Trichophyton interdigitale and Microsporum quinckaenum.[29] Galactofuranosyl residues occurred as nonreducing termini. The galactomannans were fractionated into galactomannans I and II. In general, galactomannan I contained less galactose than galactomannan II.

Aspergillus fumigatus, the organism responsible for aspergillosis, produces a peptidogalactomannan which contains galactofuranosyl residues.[30,31] The polypeptide region

is rich in seryl, theronyl, and alanyl residues and contains only traces of tyrosine, cysteine, and methionine.[10,30]
The mannopyranosyl residues are attached to one another by
(1→2) and (1→6) α-linkages. Galactomannans obtained from
either the growth medium or mycelia had a mass of 20,000 to
30,000 daltons and were homogenous by ultracentrifugal or
electrophoretic analysis. A ratio of galactose:mannose of
1:1 was obtained. Galactomannans were also obtained
Aspergillus flavus, Aspergillus effsus, Penicillium charlesii,
Pencillium notatum, Pencillium frequentans, and Penicillium
expansum by Azuma et al.[10] They conclude that galactomannans
are common constituents of Aspergillus and Penicillium
species. Galactomannan from culture filtrates of Penicillium
chrysogenum contain galactofuranosyl residues and phosphorus.[16]

Lloyd[8,9] examined the "peptido-polysaccharides" from
Cladosporium werneckii and found polymers of mass 150,000
to 200,000 daltons, composed of 11 percent protein, mannosyl-containing oligosaccharides and phosphogalactomannan
attached to the polypeptide through seryl and threonyl
linkages. The peptide is rich in seryl and threonyl residues and it contains only traces of tyrosine, phenylalanine,
methionine and cysteine. Methylation analyses showed that
mannopyranosyl residues are attached primarily by (1→2) links.
Smaller quantities of (1→6) and (1→3) links also occur. The
phosphogalactomannan contains phosphodiester groups which
bridge a side chain of seven mannopyranosyl and one galactofuranosyl residues to the main core of polysaccharide. The
main core also contains 7 phosphodiesters which bridge about
33 hexosyl residues. Phosphogalactomannan contained both
galactofuranosyl and galactopyranosyl residues.

Dow and Callow[15] recently reported the occurrence of
glycopeptides in culture filtrates of Fulvia fulva (Cooke)
Ciferri (syn. Cladosporium fulvum) which contained major
quantities of D-galactofuranosyl, D-glucosyl, and D-man-
nosyl residues and smaller amounts of D-glucuronosyl,
D-galacto-saminyl and D-glucosaminyl residues. Phosphodiesters bridged glucosyl-, galactosyl- and mannosyl-containing side chains to the main core of polysaccharide.
Bridging of side chain oligosaccharides to main polysaccharide chain through a phosphodiester which is attached to
C-1 of the reducing terminal sugar of the oligosaccharide
and to C-6 of a hexosyl residue in the main chain occurs

frequently in polysaccharides (mannans and galactomannans) obtained from fungi and yeasts.

It was suggested that the saccharides were responsible for determining the specificity in gene-for-gene interaction between Fulvia fulva (pathogen) and the host tomato plant.[32,33] However, Dow and Callow reported[34] that the leakage of electrolytes that occurred when the glycopeptide was administered to tomato leaf surfaces showed no race or cultivar specificity. The glycopeptides appeared to bind to the surface of the leaf, possibly to the plasmalemma surface.

Work on the biosynthesis of "galactocarolose" in our laboratory soon established that galactocarlose and mannocarolose in 28-day stationary cultures of Penicillium charlesii[2] were degradation products of more complex polymers which contained galactofuranosyl, mannopyranosyl, and glucopyranosyl residues.[7, 35, 36] These polysaccharide-containing polymers, secreted into the growth medium, were shown to contain phosphorus also.[35] The major polymer was released from DEAE-Sephadex-borate with 0.1 N HCl-0.06 M LiCl and was shown to be heterogenous with respect to mass, with respect to ratio of galactose:mannose:glucose, and with respect to ratio of P:galactose but not with respect to P:mannose.[36] Treatment of this polymer with 70 percent formic acid at 100°C for 90 minutes resulted in the release of phosphorus-containing substances that chromatographed like glucose 6-phosphate, mannose 6-phosphate and inorganic orthophosphate.[35] Later it was established that this exocellular polymer was a glycopeptide[4] and that the maximum quantity of polymer occurred in the medium within 8 to 10 days when the culture of Penicillium was grown with vigorous aeration.[6] Under these conditions little or no glucose was found in the polymer.[3] Treatment with alkali effected β-elimination of the saccharides from the polypeptide(s).[4] This treatment revealed that a phosphogalactomannan and 10 to 12 mannosyl-containing low molecular weight saccharides are attached through seryl and threonyl residues to the polypeptide(s). The mannosyl-containing saccharides released were primarily mannose, mannobiose, and mannotriose. Methylation analyses of the glycopeptide and phosphogalactomannan showed that the mannopyranosyl residues were attached by (1→2) and (1→6) α-linkages.[3] Acetolysis of the glyco-

peptide and phosphogalactomannan released mannotetraose,
mannotriose, mannobiose, and mannose in ratios of 11:5:13:16
and 11:4:5:13, respectively. These data show that 1:8:3
residues of mannotriose, mannobiose, and mannose, respectively, were released from the polypeptide during β-elimination. We also noted a phosphorus-containing saccharide
that eluted in the position of a mannopentaose.[24] This
substance has not been characterized further, but it most
likely is not a phosphomannopentaose as the phosphate would
modify the elution position of saccharides.

Phosphogalactomannan from Penicillium charlesii glycopeptide is comprised of a phosphomannan backbone to which
approximately 10 galactan chains are attached, each by (1→3)
β-linkage.[3,5,37] The galactan chains contain an average of
22 5-O-β-D-galactofuranosyl residues when the glycopeptide
is obtained from cultures prior to depletion of glucose.[37]
The exocellular glycopeptide from these cultures has a mass
of about 70,000 daltons. Glycopeptide isolated from cultures which are depleted of glucose contain an average of
two galactofuranosyl residues per galactan chain and have a
mass of about 22,000 daltons. Partial degradation of the
galactan chains results from an exo-β-D-galactofuranosidase
which is secreted into the growth medium as the medium becomes depleted of glucose,[37] (Pletcher, Lomar and Gander,
in press). This enzyme degrades the galactan rapidly until
two galactofuranosyl residues remain per chain. These two
galactofuranosyl residues are removed at least two orders of
magnitude more slowly than those from the longer chain
galactans. Carbon-13 nmr spectroscopy shows a major
resonance signal at chemical shifts of 110.5 ppm, the chemical shift of the anomeric carbon of nonreducing terminal
β-D-galactofuranosyl residue and no signal at 104 ppm which
would be indicative of a nonreducing terminal α-D-galacto-
furanosyl residue.[5] Thus, we conclude that the slow removal
of the final two galactofuranosyl residues from the mannan
must be due to steric restraints. However, an exocellular
"polysaccharide" from 28-day cultures of Penicillium varians
contains both α- and β-linked D-galactofuranosyl residues.[38]
We have obtained a glycopeptide from Penicillium varians
which has a similar hexose composition (galactose, glucose
and mannose) and have shown that the polysaccharide is attached to polypeptide as occurs in Penicillium charlesii
(Gander and Unkefer, unpublished).

During the course of studies to determine the N-terminal amino acyl residue(s) of the glycopeptide, 2-aminoethanol was found as a constituent and shown to occur in a molar ratio of 2-aminoethanol:glycopeptide of about 1:1.[4] This work was extended to show that the glycopeptide also contained 2-dimethylaminoethanol[24] and that 2-aminoethanol and 2-dimethylaminoethanol were probably attached to the mannan through phosphodiester bridges.[5] ^{31}P nmr spectra are consistent with this interpretation of the data and suggest that some 2-methyl aminoethanol and choline also occur

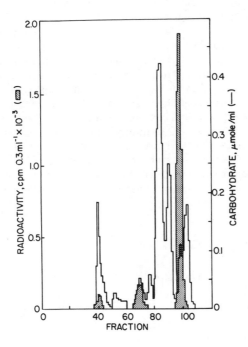

Figure 1. Distribution of ^{14}C and carbohydrate following chromatography of the low molecular weight substances from (^{14}C)ethanolamine-labeled glycopeptide on Bio-Gel P2. Glycopeptide labeled with ^{14}C from [2-^{14}C]ethanolamine was treated with alkali and fractionated on Sephadex G-50 as fractions 20 to 25 were combined, their volume was reduced to about 1 ml, and they were chromatographed on Bio-Gel P2 column. Fractions 35 to 110 were analyzed for ^{14}C (▨) and carbohydrate (□).[24]

in the glycopeptide.[39]   There is some evidence for the attachment of a neutral substance, possibly a saccharide, to
the mannan through a phosphodiester bridge (Unkefer and
Gander, unpublished).   Treatment of ^{14}C-2-aminoethanol- or
[3]H-methyl-labeled glycopeptide with alkali released about
40 percent of the ^{14}C or ^{3}H with low molecular weight substances and the remainder was nondialyzable and was reiso-
lated with phosphogalactomannan.[24]   In contrast, 95 percent
of the phosphorus remained with phosphogalactomannan following treatment of glycopeptide with alkali.   The phosphate
which was released with low molecular weight substances
eluted with "mannopentaose" fraction from BioGel P-2 column
as did a small fraction of the ^{14}C-2-aminoethanol and [3]H-
methyl-labeled substances.   The remainder of the [14]C- and
[3]H-labeled substances eluted near the monosaccharides (Fig. 1).
Initially we concluded that the phosphodiesters were in two
environments; the alkali stable phosphodiesters being attached to C-6 hydroxymethyl groups of the mannan and the
alkali unstable phosphodiesters being attached to alkali
unstable C-2 position of mannopyranosyl residues.[24]   Preliminary evidence shows that alkali releases choline, probably because of its neutral charge even in 0.4 N alkali
which is in contrast to 2-aminoethanol and its mono- and
dimethylamino derivatives (Unkefer and Gander, unpublished).

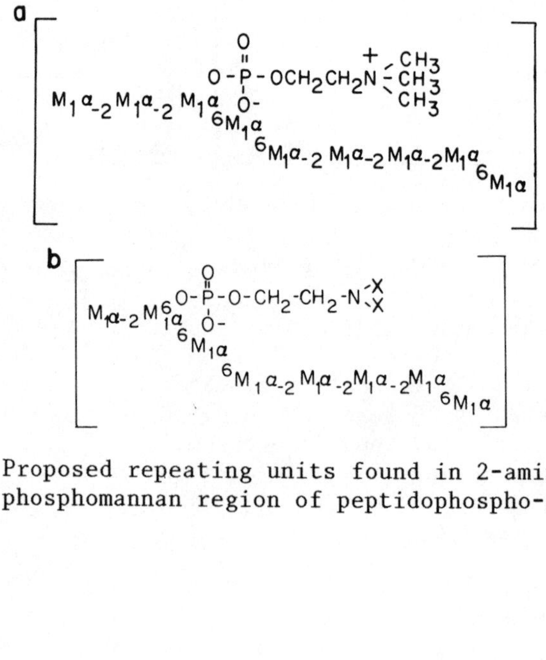

Figure 2.   Proposed repeating units found in 2-aminoethanol-
containing phosphomannan region of peptidophospho-galacto-
mannan.

There was no evidence, based on ^{31}P NMR spectra, for either
phosphomonoester or cyclic phosphate occuring in the mannan
after treatment with 0.4 N alkali. A model which is consistent with the data obtained is shown in Figure 2. This is
but one of several feasible models and it assumes that
phosphogalactomannan is composed of two types of repeating
units; one with phospho-2-aminoethanol, or its mono- or
dimethylamino derivative, bridged to C-6 of the reducing
terminal residue of a mannooctaose (Fig. 2b) and the other
with phospho-choline bridged to the C-6 of a mannononaose
region of the phosphogalactomannan. As yet, we do not know
to what region of phosphogalactomannan the neutral phosphodiester is attached. The single phospho-2-dimethyl-aminoe-
thanol- and mannosyl-containing oligosaccharide which is
released by treatment with alkali[24] may serve as a "stop"
signal during biosynthesis of phosphomannan or phosphogalactomannan. Drewes and Gander[40] observed that a mutant
of Penicillium charlesii which was selected because of its
inability to grow on D-galactose, produced exocellular
glycopeptide of about 23,000 daltons which contained only
one mole of phosphorus and two to three moles of galactose
per mole of glycopeptide. This suggests that phosphoryl
residues are attached only after galactofuranosyl residues
are attached. The occurrence of the one phosphoryl residue
may represent that incorporated as the "stop" signal during
synthesis of the mannan. The data also show that in the
mutant the mannan can be synthesized without attachment of
the galactan or phospho-2-aminoethanol derivative. It is
not known whether the two or three galactofuranosyl residues
are attached to the region of the mannan proposed as the
stop signal or to some other region.

In vitro studies on the biosynthesis of the mannose-
containing low molecular weight saccharides have established that GDP-mannose serves as the mannosyl donor in formation of mannobiosyl-peptidophosphogalactomannan from man-
nosyl-peptidophosphogalactomannan.[41] However, there was no
evidence for formation of mannotriosyl or mannotetraosyl
residues. A system which incorporates mannosyl residue(s)
into phosphogalactomannan region was separated from that
which incorporates mannosyl residues into the low molecular
weight saccharides.[42] This system resulted in adding
(1→6)-linked and (1→2)-linked mannosyl residues to phosphogalactomannan.

A peptidophosphogalactomannan-dependent incorporation of 2-aminoethanol from CDP-2-aminoethanol is catalyzed by membrane-bound enzymes.[43] Treatment of the isolated peptidophosphogalactomannan containing ^{14}C-2-aminoethanol with alkali resulted in the release of about 40 percent of the 2-aminoethanol with the low molecular weight substances and the remainder of the 2-aminoethanol was associated with phosphogalactomannan.

Investigations on the composition and sequence of amino acyl residues in the polypeptide region of the glycopeptide have shown that it contains 8 seryl, 7 threonyl, 4 alanyl, 2 valyl, 3 glycyl, 2 prolyl, 1-2 glutamyl-glutaminyl, and 1 aspartyl-asparignyl residues and less than one residue each of histidine, leucine, isoleucine, and lysine.[4] The polypeptide has negligible quantities of S-containing and aromatic amino acyl residues. Treatment of the glycopeptide with alkali released carbohydrate from 6 of the seryl residues and 7 of the threonyl residues. The molecular weight of the polypeptide is about 3,000 and was shown to be nearly homogenous by gel filtration chromatography.[44] Analyses for N-terminal amino acyl residues showed that the glycopeptide contained N-terminal seryl, aspartyl, glycyl and glutamyl-glutaminyl residues.[4] Although treatment of the glycopeptide with pronase released one to three amino acyl residues[44] this treatment did not eliminate the heterogeneity at either C-terminal or N-terminal ends of the polypeptide (Tonn and Gander, unpublished). Treatment of the glycopeptide with anhydrous HF[45] removes the carboydrate without cleavage of the polypeptide (Tonn and Gander, unpublished). The polypeptide was fractionated into four polypeptides containing N-terminal seryl, glycyl, and aspartyl-asparaginyl residues and part of the heterogeneity with respect to the number of leucyl, isoleucyl and lysyl residues was resolved by this treatment. However, we have not been able to obtain a unique sequence for any of these polypeptides. From these studies we conclude that the exocellular glycopeptide is derived from several different glycoproteins. The role of the phosphogalactomannan may be to provide the proper recognition for transport and exocytosis of these proteins.

Exocellular glycopeptides were isolated and purified from the following species of Penicillium: notatum, chrysogenum, patulum, claviforme, raistrickii and

roqueforti. Each glycopeptide was shown to have an amino
acyl composition similar but not identical to that of
Penicillium charlesii. Carbon-13 nmr spectroscopy of
glycopeptides containing natural abundance ^{13}C showed that
each glycopeptide had large quantities of 5-0-β-$\underline{D}$-galacto-
furanosyl residues (Gander and Rees, unpublished). Thus it
is apparent that 5-0-β-$\underline{D}$-galactofuranosyl-containing glycopeptides are common among the Ascomycetes.

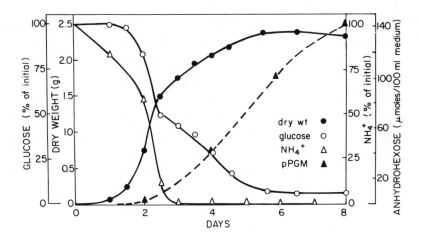

Figure 3. Time course of $\underline{D}$-glucose and NH_4^+ uptake from
the growth medium and increase in dry weight of mycelia of
P. charlesii cultures, and the release of peptidophosphogalactomannan, pPGM, into the medium. The organism was
cultured in a modified Raulin-Thom medium (Jordan, J. M.
and J. E. Gander. 1966. Biochem. J. 100:694-701). The
contents of the flasks were removed at the indicated inter-
vals, filtered, and the filtrate was assayed for the total
carbohydrate (o-o), NH_4^+ (Δ-Δ), dry weight (●-●) and pPGM
(▲-▲). The mycelia were dried for 24 hours at 80°C and
weighed. The initial concentrations of $\underline{D}$-glucose and NH_4^+
were 278 and 36.3 mM, respectively.

PARTIAL CHARACTERIZATION OF CELLULAR PEPTIDOPHOSPHO-
GALACTOMANNAN

No galactofuranosyl-containing substances were found
in the growth medium before 2.5-3 days (Fig. 3).[6] However,
radioactive exocellular glycopeptide was isolated from
9-day cultures when Penicillium charlesii was cultured for
2.5 days in ^{32}P inorganic phosphate, D̲-[^{14}C]glucose, or
L̲-[^{14}C]-threonine and the culure transferred into nonradio-
active medium which had supported Penicillium charlesii for
2.5 days. The quantity of radioactivity incorporated suggested that glycopeptide synthesis started at the time, or
soon after, germination.

Three day cultures of Penicillium charlesii were
washed, the mycelia were broken with Al_2O_3, and the cell
walls and unbroken cells were separated from membranes and
soluble-cytoplasmic fraction by ultracentrifugation. Both
the membranes and the cytoplasmic fraction contained
galactofuranosyl residues which were released by exo-β-D̲-
galactofuranosidase.[46] Glycopeptides in the soluble-cytoplasmic fraction were isolated by procedures routinely used
in this laboratory[3] (Fig. 4). A galactofuranosyl-containing
glycopeptide was obtained. The glycopeptide was somewhat
larger in that it contained approximately 60 amino acyl
residues and a mass of 80,000 to 90,000 daltons as determined by gel filtration chromatography and SDS disc polyacrylamide gel electrophoresis.[46] Treatment of the glycopeptide with alkali released mannosyl, mannobiosyl. and a
galactofuranosyl-containing phosphogalactomannan. The
glycopeptide contained N-terminal seryl and glycyl residues
and dansyl-2-aminoethanol was found following treatment of
the glycopeptide with 1-dimethylamino naphthalene 5-sulfo-
chloride (dansyl chloride) and hydrolysis in acid. The
polypeptide contained approximately 16 and 11 seryl and
threonyl residues, respectively, and only traces of S-containing or aromatic amino acids were found. Polypeptide
accounts for only 14 percent of the polymer. Some of the
chemical properties of the glycopeptides obtained in this
experiment are shown in Table 1. The small, but finite,
quantity of the glycopeptide in the growth medium at 3 days
is consistent with the data obtained previously.[6]

The ratio of galactofuranosyl:mannosyl residues is 2:1
which is typical of that observed previously[37] in glycopeptide

from culture filtrates prior to the release of large quanti-
ties of exo-β-$\underline{\underline{D}}$-galactofuranosidase. Based on our previous
studies[3,7] we anticipated a molar ratio of hexosyl resi-
dues:P of 30:1. The observed ratio of 15:1 suggests that
each mole of glycopeptide contains 20 moles of P. We note
that the sum of moles of mannotetraosyl, mannotriosyl, and
mannobiosyl residues per mole of P derived from phosphogalactomannan by actolysis is 1:1. This suggests that each

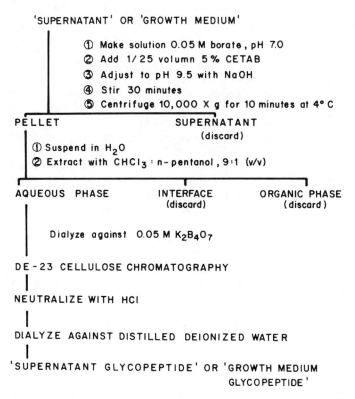

'SUPERNATANT' OR 'GROWTH MEDIUM'

    ① Make solution 0.05 M borate, pH 7.0
    ② Add 1/25 volumn 5% CETAB
    ③ Adjust to pH 9.5 with NaOH
    ④ Stir 30 minutes
    ⑤ Centrifuge 10,000 X g for 10 minutes at 4° C

PELLET   SUPERNATANT
                         (discard)
① Suspend in H_2O
② Extract with $CHCl_3$: n-pentanol, 9:1 (v/v)

AQUEOUS PHASE   INTERFACE   ORGANIC PHASE
                 (discard)   (discard)

    Dialyze against 0.05 M $K_2B_4O_7$

DE-23 CELLULOSE CHROMATOGRAPHY

NEUTRALIZE WITH HCl

DIALYZE AGAINST DISTILLED DEIONIZED WATER

'SUPERNATANT GLYCOPEPTIDE' OR 'GROWTH MEDIUM
                             GLYCOPEPTIDE'

Figure 4. Flow diagram for isolation of soluble glycopeptide obtained from the soluble cytoplasmic fraction.

Table 1. Chemical characterization of glycopeptides

Source of Glycopeptide	μmoles An-hex^a Flask	An-hex^a Phosphate^b	An-hex^a Protein	Galactose An-hex^a
Growth Medium	2.5	16	6.7	0.66
Supernatant	10.4	15	7.0	0.64
DOC-soluble	4.0	12	3.8	0.60

^{a}Anhydrohexose.

^{b}Polymer treated with 0.05 N HCl for 90 minutes at 110°C.
Galactose released measured by the coupled galactose
oxidase-horseradish peroxidase assay.

region containing α-1-2-linked mannopyranosyl residues may
contain a phosphodiester attached to it. Recent studies
using ^{31}P nmr spectroscopy to characterize the glycopeptide
and phosphogalactomannan suggest that glycopeptides isolated
from culture filtrates of 6-day Penicillium charlesii may
contain P in an environmenment not represented by those
reported previously (Unkefer and Gander, unpublished).
Using data from Table 1, molar ratios of P:glycopeptide of
20:1, 20:1 and 22:1 were obtained for exocellular glycopeptide, soluble cytoplasmic glycopeptide and membrane-
bound glycopeptide, respectively.

By all of the criteria which we have used, the glycopeptide obtained in the fraction containing soluble cytoplasmic substances appears to be a precursor to the exocellular polymer. Its major difference is in having about
twice the number of amino acyl residues.

PARTIAL CHARACTERIZATION OF MEMBRANE-BOUND LIPO-PEPTIDO-
PHOPHOSGALACTOMANNAN

The membranes obtained by centrifuging the cell-free,
wall-free homogenate described above were fractionated by
isopycnic sucrose gradient ultracentrifugation. Six bands
of membranes were observed and the occurrence of galactofuranosyl residues and GDP-D-mannose mannosyltransferase in

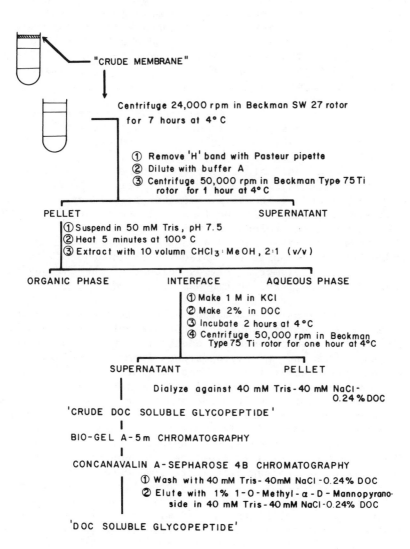

Figure 5.   Flow diagram for isolation of lipo-glycopeptide
obtained from membrane fraction V, ρ = 1.18 g/cc.

each was determined.[46] Mannosyltransferase activity was
located primarily in the low density membranes and galactofuranosyl residues were located in membranes of density
1.18 g/cc. The galactofuranosyl-containing substances were
solubilized with KCl:deoxycholate and purified by chromatography on BioGel A 5m followed by chromatography on concanavalin A-Sepharose 4B[46] (Fig. 5). The galactofuranosyl-
containing polymer was amphipathic.

The composition of this polymer was determined and it
was found to contain 26 percent protein, a galactofuranosyl:
mannopyranosyl ratio of 3:2, an estimated mass of about
105,000 daltons, an amino acyl composition similar to that
of exocellular and soluble glycopeptide except seryl and
threonyl residues, 24 and 19, respectively, account for
only 38 percent of the total amino acyl residues. Treatment
with alkali resulted in the loss of 11 seryl and 9 threonyl
residues with the release of primarily mannobiose. No
phosphogalactomannan was obtained. This suggests that the
hydrophobic portion of the molecule was attached to this
region of the polymer. Only 49 percent of the carbohydrate
applied to BioGel P-2 column was recovered.

Evidence that the membrane-bound glycopeptide(s) contained a hydrophobic substance not present in the soluble
glycopeptide was obtained by adding [1-^{14}C]acetate to the
growth medium and the glycopeptides from the medium,
soluble cytoplasmic fraction and membranes from 3-day cultures of Penicillium charlesii were isolated (Table 2).
Glycopeptide obtained from the growth medium and that in the
soluble-cytoplasmic fraction contained 2,000-3,000 cpm/μmole
of hexosyl residues.[46] In contrast, the membrane-bound
glycopeptide contained 35,000 cpm/μmole of hexosyl residues.
The ^{14}C was not released from the glycopeptide by alkaline
conditions used to saponify fatty acyl esters or steryl
esters. The ^{14}C was released by conditions required for
releasing sphingosine bases, or their derivatives from
sphingolipids. Although the ^{14}C migrated like sphingosine
in two solvent systems when the CHCl$_3$:methanol extract
obtained following treatment of the ^{14}C-labeled glycopeptide with 4 N NaOH at 110°C for 5 hours, was subjected
to thin layer chromatography in four solvent systems, the
^{14}C migrated unlike sphingosine in the other two solvent
systems. Treatment of the ^{14}C-containing substances with
IO$_4$- resulted in the formation of a product which, based on

Table 2. ^{14}C from [1-^{14}C]acetate incorporated into
deoxycholate-soluble glycopeptide

Source of glycopeptide[a]	^{14}C/μmole Anhydrohexose
	cpm x 10^{-3}
Growth medium	1.9
Supernatant	2.7
Deoxycholate-soluble	35.3

[a]Glycopeptides were isolated and purified from cultures
that were grown in four flasks under standard conditions.
Twenty-four hours after inoculation 1 mCi [1-^{14}C]acetic
acid (58 mCi/mmole) was added to each of two flasks.

preliminary analyses by gas chromatography-mass spectro-
metry, suggests that the base is phytosphingosine or dehy-
drophingosine. Dehydrosphingosine occurs to the extent of
3 percent of the sphingolipid bases in yeasts.[47]

     The position of attachment of the hydrophobic moiety
or moieties to the phosphogalactomannan is unknown, nor do
we know what signal serves to release the peptidophosphogalactomannan from lipo-peptidophosphogalactomannan. The
occurance of lipo-peptidophosphogalactomannan seems to be
restricted to membranes with a density of about 1.2 g/cc.[46]
The organelles from which these membranes are derived are
unknown, but it seems unlikely that they are either endoplasmic reticulum or Golgi as they have little or no GDP-D-
mannose mannosyl transferase activity.   Schibeci et al.[47]
fractionated plasma membranes from yeast protoplasts into
two major bands with buoyant densities of 1.15-1.17 and
1.17-1.19 g/cc. Thus, the membranes obtained from
Penicillium charlesii[46] may be derived from plasma membrane.

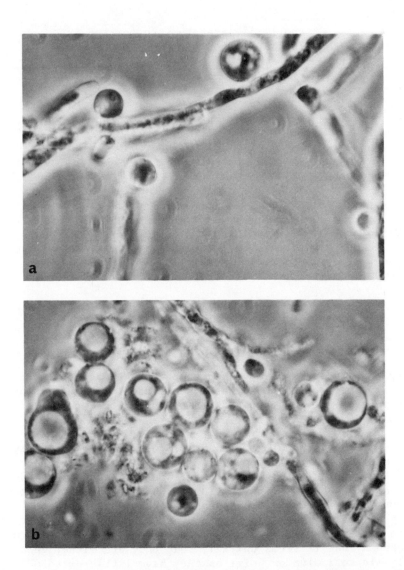

Figure 6. Penicillium charlesii hyphae and protoplasts a) 0.5 hour and b) 2.5 hours after initiaion of treatment with β-D-glucuronidase (3 mg/ml) and cellulysin (20 mg/ml) in 50 mM maleate-0.5 M KCl, pH 6.0 buffer. 980x magnification.

LOCATION OF LIPO-PEPTIDOPHOSPHOGALACTOMANNAN

Experiments were conducted to locate lipo-glycopeptide in Penicillium charlesii. Protoplasts were prepared by treating hyphae which were obtained after culturing the fungus for 42 hours, with a mixture of cellulysin (20 mg/ml) and β-D-glucuronidase (3 mg/ml) in 50 mM maleate-0.5 M KCl buffer, pH 6.0 for 2.5 hours (Laybourn and Gander, unpublished). A typical protoplast preparation obtained after treatment with cellulysin and β-D-glucuronidase for 0.5 and 2.5 hours is shown in Figure 6a and 6b, respectively. The protoplasts were washed with maleate-KCl buffer and filtered through Pellon to remove the debris (Fig. 7). A preparation was fixed with glutaraldehyde, sectioned and stained with OsO_4. The preparation contained protoplasts which were essentially free of cell walls and protoplasts which had started resynthesis of cell wall material, (spheroplasts) (Fig. 8a and 8b, respectively).

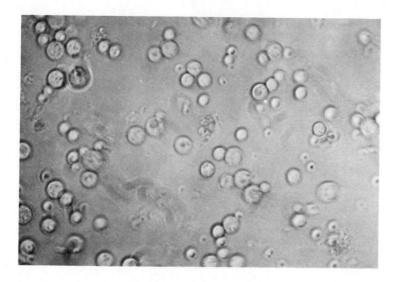

Figure 7. Protoplasts obtained from P. charlesii hyphae after removal of cell debris by filtration through medium porosity Pellon. 980x magnification.

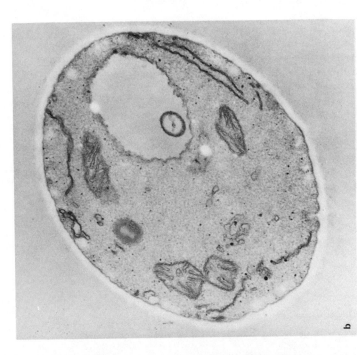

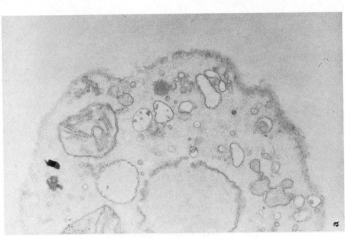

Figure 8.　Electron micrograph of a) protoplast and b) spheroplast from Penicillium charlesii hyphae; 55,000x magnification. The protoplast/spheroplast preparation was fixed with 2.5% (v/v) glutaraldehyde and 2.3% (w/v) formaldehyde in the 50 mM maleate-0.5 M KC1, pH 6.0 buffer. The protoplasts were rinsed in this buffer and post fixed in 1% (w/v) OsO$_4$ and buffer. The preparation was embedded in Spurr's (Fullam), cut with a diamond knife on a Sorvall NT2B, stained in alcoholic uranyl acetate and post-stained with lead citrate. Sections were observed with a Hitachi model HU11C electron microscope.

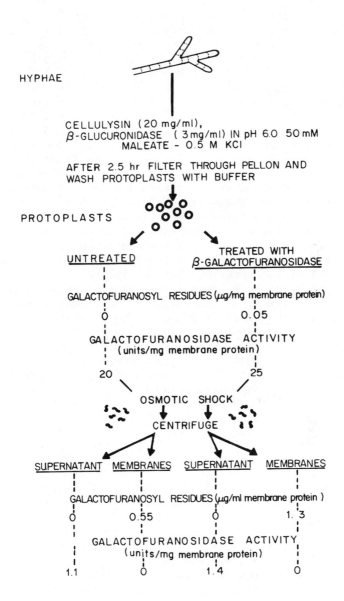

Figure 9. Distribution of lipo-glycopeptide and exo-β-$\underline{D}$ galactofuranosidase in <u>Penicillium</u> <u>charlesii</u> protoplasts.

Investigations were undertaken to locate the galactofuranosyl-containing substances and exo-β-$\underline{D}$-galactofurano-
sidase in the protoplast/spheroplast preparations. Protoplasts were either pretreated with added exo-β-$\underline{D}$-galacto-
furanosidase at pH 4 for 24 hours or held for 24 hours at
pH 4 without added exo-β-$\underline{D}$-galactofuranosidase. This
treatment released 0.05 μg of galactofuranosyl residues
from the protoplasts treated with galactofuranosidase
(Fig. 9). The protoplast preparation was washed with
acetate-KCl buffer and exo-β-$\underline{D}$-galactofuranosidase activity
was measured. Both preparations contained about 20 units
of enzyme activity (Fig. 9). Separate samples of galactofuranosidase treated and untreated protoplasts were lysed by
transferring them to distilled, deionized H_2O and the membranes were separated from the soluble cytoplasmic material
by ultracentrifugation. Membrane and soluble cytoplasmic
fractions were analyzed for galactofuranosyl residues and
for galactofuranosidase activity. Galactofuranosyl residues
were found in the membranes but not in the soluble cytoplasmic fraction. The membranes derived from protoplasts
which were pretreated with galactofuranosidase contained 1.3
μg of galactofuranosyl residues per mg of protein in the
membranes as compared with only 0.55 μg in the membranes
which were untreated. Pretreatment with galactofuranosidase
may have made the membranes more susceptible to the action
of proteases so that the relative quantity of membrane
protein per unit of membrane was less in those protoplasts
which were pretreated with galactofuranosidase . In contrast to the location of galactofuranosyl residues, galactofuranosidase activity was found only in the soluble cytoplasmic fraction (supernatant), and the activity per unit of
membrane protein was decreased considerably from that
observed in the intact membranes. This suggests that
activity was destroyed during lysis and separation of the
membranes from the soluble subtances. This destruction may
have been the result of proteolysis.

The membranes obtained following lysis were fractionated by isopycnic sucrose gradient ultracentrifugation.
Galactofuranosyl residues were found primarily in fractions
with densities of 1.21 to 1.24 g/cc and negligible quantity
was found in the lighter membranes (Fig. 10). These data
are consistent with those obtained with membranes isolated
from homogenates of mycelia.[46] We conclude that the galactofuranosyl-containing lipoglycopeptide is located at least in

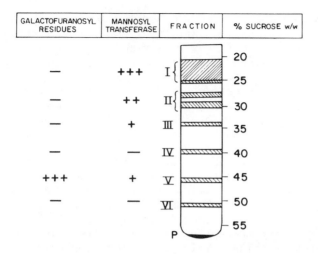

Figure 10. Separation of membranes of <u>Penicillium</u> <u>charlesii</u> protoplasts and location of galactofuranosyl residues and mannosyltransferase activity.

part on the outer surface of the plasma membrane, but that the larger portion of lipo-glycopeptide may be attached to an organelle which fuses with the plasma membrane.

PROPOSED FUNCTION OF EXOCELLULAR PEPTIDOPHOSPHOGALACTOMANNAN

    These results when coupled with those presented previously[46] show that at least two types of galacto-furanosyl-containing substances occur in <u>Penicillium</u> <u>charlesii</u>: a) a membrane-bound lipo-glycopeptide and b) soluble glycopeptides. The data suggest that lipoglyco-peptide is the precursor of the soluble glycopeptides; that is, the polypeptide constitues a greater portion of the total mass in the lipo-glycopeptide than in the soluble forms. The lack of galactofuranosyl-containing substances in the soluble-cytoplasmic fraction of protoplasts suggest that the "soluble cytoplasmic" glycopeptide obtained from mycelia[46] is not intracellular but instead represents glycopeptide released form the plasma membrane. This glycopeptide is probably that which is in transit from the surface of the membrane to the outside of the cell wall and is only loosely associated with the cell wall. During the transition of the glycopeptide from its membrane-bound form to exocellular

glycopeptide the polypeptide becomes degraded by proteases
located on the outside surface of the plasma membrane, in
the cell wall, and in the growth medium. This conclusion
is consistent with the observation that the number of amino
acyl residues in the polypeptide region decrease from about
100 in lipo-glycopeptide to 30 in exocellular glycopeptide.
The data suggest that this serves as a mechanism for de-
grading fungal glycoproteins in regions of hyphae containing
older cells, when the culture becomes deficient in N-containing nutrients. The degradation process may occur
primarily on the outer surface of the plasma membrane, a
process which serves to release amino acids out into the
medium. The released amino acids are then available to the
growing hyphal tips. The heterogeneity observed in the
polypeptide region of exocellular glycopeptide[44] suggests
that the lipo-glycopeptide is derived from many intracellular glycoproteins. This process may represent the mechanism by which Ascomycetes couple turnover of glycoproteins
to conservation of N-containing substances until these substances are needed by growing and new members of the fungal
colony. This conclusion is also consistent with the onset
of appearance of proteases in the medium.

     The role of phosphogalactomannan in this process may
be of prime importance. It may contain the necessary
recognition sites required for endocytosis by appropriate
organelle(s) of the glycoproteins to be turned over. These
organelles may fuse with the plasma membrane; a process
which would place the glycoprotein on the outside surface
of the plasma membrane.

     Currently we do not know whether the phosphogalacto-
mannan is put onto the glycoprotein as one unit or if it
is derived from mannoproteins which are modified by the
addition of galactofuranosyl and the appropriate phospho-2-
aminoethanol derivative(s). Experiments are in progress to
resolve this question.

     The exocellular glycopeptide may have an important
role in maintaining adequate minimal quantities of key
nutrients so that the Penicillium can survive when the
growth medium becomes nearly depleted of these nutrients.
The sequence of events, as they are currently known, following formation of lipo-glycopeptide and release of exocellular glycopeptide and exo-β-$\underline{D}$-galactofuranosidase are

summarized in Figure 11. Following depletion of NH_4^+ exocellular glycopeptide is released from the organism and
diffuses away. Proteases, phosphatases, and some glycosidases are also released. Following depletion of glucose,
the pH increases and organic acids are utilized by the
fungus. These conditions result in the release of exo-β-$\underline{D}$-
galactofuranosidase (Pletcher, Lomar and Gander, in press).
Degradation of the galactan region of exocullular glycopeptide provides sufficient carbohydrate for survival of
the Penicillium. The glycopeptide also contains other substances that may be critical for survival of the orgamism;
that is, substances like phosphorus, and 2-aminoethanol and
its methyl derivatives. We believe that "survivalin" is an

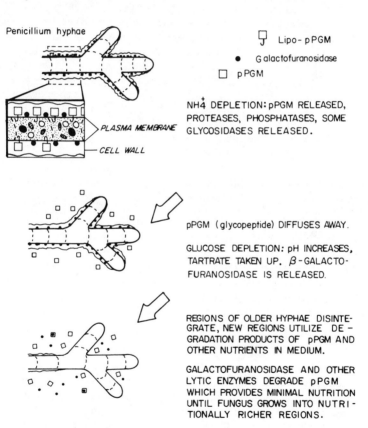

Figure 11. Model of proposed lipo-glycopeptide metabolism
in Penicillium charlesii.

appropriate descriptive name for this group of glycopeptides as they undoubtedly provide quantities of carbohydrate, N-containing substances and phosphate sufficient
to allow the organism to survive in a nutritionally deficient
environment.

     The exocellular glycopeptide may also have a role in
directing growth of Penicillium toward a new nutrient source.
The proposed roles of exo-β-D̲-galactofuranosidase and exocellular glycopeptide in directing growth on a solid surface
is summarized in Figure 12. This figure shows a sequence
of events in which the glycopeptide is released and diffuses
on the surface away from the fungal colony. As the colony
depletes available carbohydrate in its growth region,
exo-β-D̲-galactofuranosidase is released and also diffuses
over the surface. If a polymer-containing nutrient source
is located on the surface then exo-β-D̲-galactofurano-
sidase will be concentrated on the polymer if the enzyme
binds tightly to any of the polymers. Increased enzyme
activity will result in increased rate of galactose release
from exocellular glycopeptide which, in turn, will establish a gradient of galactose concentration back to the
fungal colony. The fungus will grow along with the gradient
of galactose toward the nutrient source.

     This model suggests that a portion of the specificity
between host plant and pathogenic fungus may reside with
the ability of the glycohydrolase released when the nutrient
source of carbohydrate becomes depleted, to bind to polymers
which are secreted by the plant. Dow and Callow suggest
that exocellular fungal glycopeptide from Fulvia fulva
binds to the tomato leaf and increases permeability of
phosphate which results in its release from the cell.[34]
However, binding of glycopeptide to the leaf plasma membrane
was not specific for race of pathogen or cultivar of host.

CONCLUDING REMARKS

     This chapter reviews the status of work in our labora-
tory on the structure and function of the galactofuranosyl-
and phospho-2-aminoethanol-containing fungal glycopeptides.
We suggest that the glyclopeptides are derived from a series
of glycoproteins and that these glycoproteins are, at least
in part, degraded on the outside of the plasma membrane but
only after the medium becomes essentially depleted of its

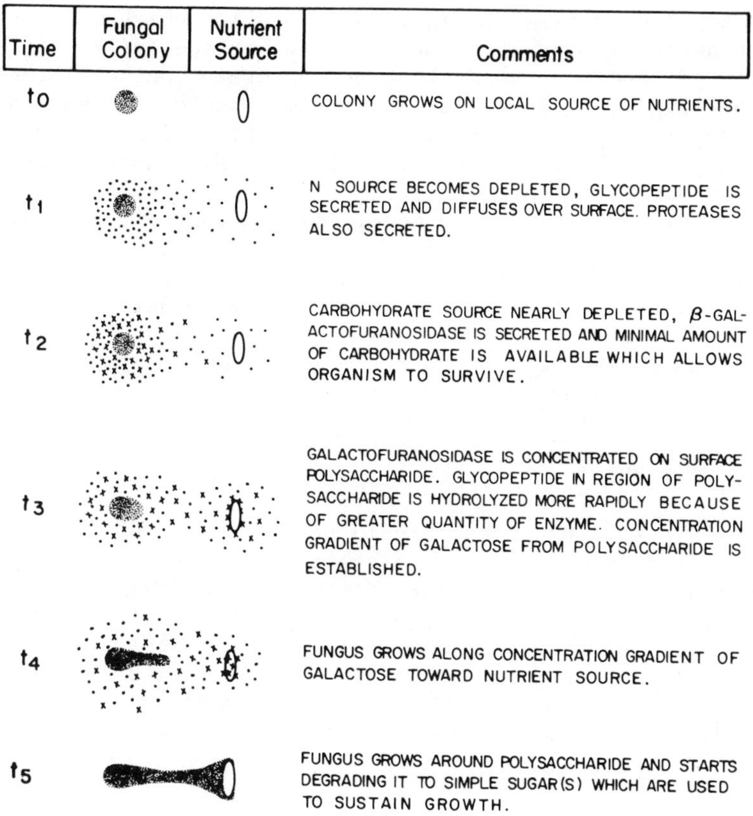

Time	Fungal Colony	Nutrient Source	Comments
t_0			COLONY GROWS ON LOCAL SOURCE OF NUTRIENTS.
t_1			N SOURCE BECOMES DEPLETED, GLYCOPEPTIDE IS SECRETED AND DIFFUSES OVER SURFACE. PROTEASES ALSO SECRETED.
t_2			CARBOHYDRATE SOURCE NEARLY DEPLETED, β-GALACTOFURANOSIDASE IS SECRETED AND MINIMAL AMOUNT OF CARBOHYDRATE IS AVAILABLE WHICH ALLOWS ORGANISM TO SURVIVE.
t_3			GALACTOFURANOSIDASE IS CONCENTRATED ON SURFACE POLYSACCHARIDE. GLYCOPEPTIDE IN REGION OF POLYSACCHARIDE IS HYDROLYZED MORE RAPIDLY BECAUSE OF GREATER QUANTITY OF ENZYME. CONCENTRATION GRADIENT OF GALACTOSE FROM POLYSACCHARIDE IS ESTABLISHED.
t_4			FUNGUS GROWS ALONG CONCENTRATION GRADIENT OF GALACTOSE TOWARD NUTRIENT SOURCE.
t_5			FUNGUS GROWS AROUND POLYSACCHARIDE AND STARTS DEGRADING IT TO SIMPLE SUGAR(S) WHICH ARE USED TO SUSTAIN GROWTH.

Figure 12. Proposed role for exocellular glycopeptide in directing growth of Penicillium.

N-containing nutrient(s). Thus the process of protein turnover is coupled to conservation of nitrogen, particularly amino acids, to allow the growing regions of the fungal colony to survive even under nutritionally stressful conditions. We have suggested that this class of glycopeptides might be called "survivalins" because of their role in the survival of the fungus when it is subjected to nutritionally stressful conditions. We also suggest three functions for the phosphogalactomannan region of the glycopeptides: a) it contains the necessary recognition sites

for the endocytosis and exocytosis process which must occur
in transporting protein into specific organelles within the
cell and exporting these lipo-glycoproteins out of the cell,
respectively, b) it provides a minimal source of amino acids,
carbohydrates and other nutrients as the medium in the region
of the colony becomes depleted of these substances, and c)
it aids in directing the growth of the organism toward a new
source of nutrients. The model will undoubtedly need to be
refined and the details of the various processes need to be
defined in terms of the molecular events that occur and the
regulation of these events. However, the general model
presented is currently serving as a working hypothesis in
our studies on the structure and function of lipopeptidophos-
phogalactomannans in Ascomycetes.

ACKNOWLEDGEMENT

This work was supported in part by Grant GM 19978 from
the Institute of General Medical Sciences, National
Institutes of Health, United States Public Health Services,
and by the Graduate School of the University of Minnesota.
University of Minnesota Agricultural Experiment Station,
Scientific Journal Series No. 11,425, Agricultural
Experiment Station, University of Minnesota,
St. Paul, Minnesota 55108.

REFERENCES

1. Clutterbuck, P. W., W. N. Haworth, H. Raistrick, G.
    Smith, and M. Stacey. 1934. XVI. Studies in the
    biochemistry of micro-organisms. 36. The metabolic
    products of Penicillium charlesii G. Smith. Biochem. J.
    28:94-110.
2. Haworth, W. N., H. Raistrick, and M. Stacey. 1937.
    LXXVI. Polysaccharides synthesized by micro-organisms.
    III. The molecular structure of galactocarolose pro-
    duced from glucose by Penicillium charlesii G. Smith.
    Biochem. J. 31:640-644.
3. Gander, J. E., N. H. Jentoft, L. R. Drewes, and P. D.
    Rick. 1974. The 5-0-β-D-galactofuranosyl-containing
    exocellular glycopeptide of Penicillium charlesii.
    Characterization of the phosphogalactomannan . J.
    Biol. Chem. 249:2063-2072.

4. Rick, P. D., L. R. Drewes, and J. E. Gander. 1974.
The 5-O-β-D-galactofuranosyl-containing exocellular
glycopeptide from Penicillium charlesii. Occurrence of ethanolamine and partial characterization of
the peptide portion and the carbohydrate-peptide linkage.
J. Biol. Chem. 249:2073-2078.
5. Unkefer, C. J., and J. E. Gander. 1979. The
5-O-β-D-galactofuranosyl-containing glycopeptide from
Penicillium charlesii. Carbon-13 nuclear magnetic
resonance studies. J. Biol. Chem. 254:12131-12135.
6. Drewes, L. R., P. D. Rick, and J. E. Gander. 1975.
In vivo biosynthesis of peptidophosphogalactomannans
in Penicillium charlesii. Arch. Microbiol. 104:101-104.
7. Preston, J. F. and J. E. Gander. 1968. Isolation and
partial characterization of the extracellular
polysaccharides of Penicillium charlesii. I. Occurrence of galactofuranose in high molecular weight
polymers. Arch. Biochem. Biophys. 124:504-512.
8. Lloyd, K. O. 1970. Isolation, characterization, and
partial structure of peptido-galactomannans from the
yeast form of Cladosporium weneckii. Biochemistry
9:3446-3453.
9. Lloyd, K. O. 1972. Molecular organization of a covalent peptido-phosphopolysaccharide complex from the
yeast of Cladosporium werneckii. Biochemistry
11:3884-3890.
10. Azuma, I., H. Kimura, F. Hirao, E. Tsubura, Y. Yamamura,
and A. Misaki. 1971. Biochemical and immunochemical
properties of glycopeptides obtained from Aspergillus
fumigatus. Japan J. Microbiol. 15:237-246.
11. Sakaguchi, O., M. Suzuki, and K. Yokota. 1968. Effect
of partial acid hydrolysis on precipitin activity of
Aspergillus fumigatus galactomannan. Japan J. Microbiol.
12:123-124.
12. Suzuki, S., M. Suzuki, K. Yokota, H. Sunayama, and O.
Sakaguchi. 1967. On the immunochemical and biochemical
studies of fungi. XI. Cross reaction of the polysaccharides of Aspergillus fumigatus, Candida albicans,
Saccharomyces cerevisiae and Trichophyton rubrum against
Candida albicans and Saccharomyces cerevisiae antisera.
Japan J. Microbiol. 11:269-273.
13. Barker, S. A., C. N. D. Cruickshank, and J. H. Holden.
1963. Structure of a galactomannan-peptide allergen
from Tichophyton mentagrophytes. Biochem. Biophys.
Acta 74:239-246.

14. Bishop, C. T., F. Blank, and M. Hranisavljevic-
    Jakovljevic. 1962. The water-soluble polysaccharides
    of Dermatophytes. I. A galactomannan from Trichophyton
    granulosum. Can. J. Chem. 40:1815-1825.
15. Dow, J. M., and J. A. Callow. 1979. Partial
    characterization of glycopeptides from culture fil-
    trates of Fulvia fulva (Cooke) Ciferri (syn.
    Cladosporium fulvum), the tomato leaf mould pathogen.
    J. Gen. Microbiol. 113:57-66.
16. Miyazaki, T., and T. Yadomae. 1968. Isolation of a
    water-soluble polysaccharide from the mycelium of
    Penicillium chrysogenum. Studies on fungal polysac-
    charides. Chem. Pharm. Bull. 16:1721-1725.
17. Preston, J. F., E. Lapis, and J. E. Gander. 1970.
    Immunological investigation of Penicillium. I.
    Serological reactivities of exocellular polysaccharides
    produced by six Penicillium species. Can. J. Microbiol.
    16:687-694.
18. Trejo, A. G., J. W. Haddock, G. Chittenden, and J.
    Baddiley. 1971. The biosynthesis of galactofuranosyl
    residues in galactocarolose. Biochem. J. 122:49-57.
19. Gander, J. E. and F. Fang. 1976. The occurrence of
    ethanolamine and galactofuranosyl residues attached to
    Penicillium charlesii cell wall saccharides. Biochem.
    Biophys. Res. Commun. 71:719-725.
20. Bartnicki-Garcia, S. 1970. Cell wall composition and
    other biochemical markers in fungal phylogeny. In
    Phytochemical Phylogeny, (J. Harbone, ed.). Academic
    Press, London. pp. 81-102.
21. Mahadevan, P. R., and E. L. Tatum. 1965. Relationship
    of the major constituents of the Neurospora crassa cell
    wall to wild type and colonial morphology. J. Bacteriol.
    90:1073-1081.
22. Mahadevan, P. R., and E. L. Tatum. 1967. Localization
    of structural polymers in the cell wall of Neurospora
    crassa. J. Cell Biol. 35:295-302.
23. Hunsley, D., and J. H. Burnett. 1970. The
    ultrastructural architecture of the walls of some
    hyphal fungi. J. Gen. Microbiol. 62:203-218.
24. Gander, J. E. 1977. The occurrence of
    N,N'-dimethylethanolamine in the 5-O-β-D-galacto-
    furanosyl-containing exocellular glycopeptide of
    Penicillium charlesii. Exp. Mycol. 1:1-8.

25. Cruickshank, C. N. D., M. D. Trotter, and S. R. Wood.
     1960. Studies on Trichophytin sensitivity. J. Invest.
     Derm. 35:219-223.
26. Wood, S. R. and C. N. D. Cruickshank. 1962. The
     relation between Trichophytin sensitivity and fungal
     infection. Brit. J. Derm. 74:329-336.
27. Baker, J. A., C. N. D. Cruickshank, J. H. Morris, and
     S. R. Wood. 1962. The isolation of trichophytin
     glycopeptide and its structure in relation to the
     immediate and delayed reactions. Immunology 5:627-632.
28. Blank, F., and M. B. Perry. 1964. The water-soluble
     polysaccharides of Dermatophytes. III. A galactomannan
     from Trichophyton interdigitale. Can. J. Chem.
     42:2862-2871.
29. Bishop, C. T., M. B. Perry, F. Blank, and F. P. Cooper.
     1965. The water-soluble polysaccharides of
     Dermatophytes. IV. Galactomannans I from Trichophyton
     gramulosum, Trichophyton interdigitale, Microsporum
     quinckeanum, Trichophyton rubrum, and Trichophyton
     schönlenii. Can. J. Chem. 43:30-39.
30. Axuma, I., H. Kimura, F. Hirao, E. Tsubura, and Y.
     Yamamura. 1967. Biochemical and immunochemical studies
     on Aspergillus. I. Chemical and biological
     investigations of lipopolysaccharide, protein and
     polysaccharide fractions isolated from Aspergillus
     fumigatus. Japan J. Med. Mycol. 8:210-220.
31. Sakaguchi, O., K. Yokata, and M. Suzuki. 1967.
     Biochemical and immunochemical studies on fungi.
     XII. On the galactomannan obtained from culture filtrate
     and cells of Aspergillus fumigatus. Yakugaku Zasshi.
     87:1268-1272.
32. van Dijkman, A., and A. K. Sijpesteijn. 1971. A
     biochemical mechanism for the gene-for-gene rsistance
     of tomato to Caldosporium fulvum. Netherlands J. Plant
     Pathol. 77:14-24.
33. van Dijkman, A., and A. Kaars Sijpesteijn. 1973.
     Leakage of pre-absorbed ^{32}P from tomato leaf discs
     infiltrated with high molecular weight products of
     imcompatible races of Cladosporium fulvum. Physiol.
     Plant Pathol. 3:57-67.
34. Dow, J. M. and J. A. Callow. 1979. Leakage of
     electolytes from isolated leaf mesophyll cells of
     tomato induced by glycoproteins from culture filtrates
     of Fulvia fulva (Cooke) Ciferri (syn. Cladosporium
     fulvum). Physiol. Plant Pathol. 15:27-34.

35. Preston, J. F., E. Lapis, S. Westerhouse, and J. E.
     Gander. 1969. Isolation and partial characterization
     of the exocellular polysaccharides of Penicllium
     charlesii. II. The occurrence of phosphate groups in
     high molecular weight polysaccharides. Arch. Biochem.
     Biophys. 134:316-323.
36. Preston, J. F., E. Lapis, and J. E. Gander. 1969.
     Isolation and partial characterization or the exocel-
     lular polysaccharides of Penicillium charlesii. III.
     Heterogeneity in size and composition of high molecular
     weight exocellular polysaccharides . Arch. Biochem.
     Biophys. 134:324-334.
37. Rietschel-Berst, M., N. H. Jentoft, P. D. Rick,
     C. Pletcher, F. Fang, and J. E. Gander. 1977.
     Extracellular exo-β-D-galactofuranosidase from
     Penicillium charlesii. Isolation, purification and
     properties. J. Biol. Chem. 252:3219-3226.
38. Jansson, P. E. and B. Lindberg. 1980. Structural
     studies of varianose. Carbohyd. Res. 82:97-102.
39. Unkefer, C. J., and J. E. Gander. 1980. Structural
     studies on the 5-0-β-D-galactoguranosyl-containing
     exocellular glycopeptide of Pencillium charlesii using
     phosphorus-31 nmr spectroscopy. Fed. Proc. 39:1634.
40. Drewes, L. R. and J. E. Gander. 1975. Exocellular
     glycopeptide from a Penicillium charlesii mutant in-
     capable growth on D-galactose. J. Bacteriol.
     121:675-681.
41. Gander, J. E., L. R. Drewes, F. Fang, and A. Lui.
     1977. 5-0-β-D-galactofuranosyl-containing exocellular
     glycopeptide of Penicillium charlesii. Incorporation
     of mannose from GDP-D-mannose into glycopeptide. J.
     Biol. Chem. 252:2187-2193.
42. Gander, J. E. and F. Fang. 1977. Properties of
     Pencillium GDP-D-mannose:glycopeptide mannosyltrans-
     ferase solubilized with Triton X-100. J. Supramol.
     Structure 6:579-589.
43. Gander, J. E. and F. Fang. 1980. Toward understanding
     the structure, biosynthesis and function of a membrane-
     bound fungal glycopeptide. Biosynthetic studies. In
     Fungal Polysaccharides (P. Sandford and K. Matsuda,
     eds.) ACS Symp. Ser. No.126, pp. 35-48.
44. Tonn, S. J. and J. E. Gander. 1977. Partial
     characterization of the peptide portion of the exocel-
     lular peptidophosphogalactomannans of P. charlesii.
     173rd Annu. Meeting, Amer. Chem. Soc., Chicago, Illinois.

45. Mort, A. and D. T. A. Lamport. 1977. Anhydrous hydrogen fluoride deglycosylates glycoproteins. Anal. Biochem. 82:289-309.
46. Gander, J. E., Beachy, J., C. J. Unkefer, and S. J. Tonn. 1980. Toward understanding the structure, biosynthesis and function of a membrane-bound glycopeptide. Structural studies. In Fungal Polysaccharides (P. Sandford and K. Matsuda, eds.). ACS Symp. Ser. No. 126, pp. 49-79.
47. Schibeci, A., J. B. M. Rattray, and D. K. Kidby. 1973. Isolation and identification of yeast plasma membrane. Biochem. Biophys. Acta 311:15-25.

Chapter Five

ENZYMATIC PROPERTIES OF PHYTOHEMAGGLUTININS

LELAND M. SHANNON AND CHARLES N. HANKINS

Department of Biochemistry
University of California
Riverside, CA   92521

INTRODUCTION

   Carbohydrate binding proteins that interact with animal
cells and agglutinate them (hemagglutinins, lectins) have
been under study for almost 100 years.[1]   Best characterized
are those hemagglutinins which have been isolated from plant
species, particularly those isolated from species of the
Leguminosae. Although hundreds of legume species contain
hemagglutinins[2] and many have been purified and characterized,[3] the physiological function(s) of these proteins is
unknown. Since the relationships among hemagglutinins from
different legume species have only recently come under careful scrutiny, functional classification of these proteins
is not possible at this time.

   The recent discovery[4,5] that certain legume enzymes
possess hemagglutinin properties opened an exciting area of

research with new avenues of approach to the general study
of plant hemagglutinins. Here we summarize what is currently known about enzymatic-hemagglutinins, how they are
related to the "classic" legume hemagglutinins and, importantly, how the study of these proteins has led to new,
useful approaches for the study of hemagglutinins. The literature of hemagglutinins has been reviewed in several[3,6,7]
excellent, recent articles. This information is useful in
the development of generalizations and provides an overview
of the prospective from which we approach our studies.

α-GALACTOSIDASE-HEMAGGLUTININS

Crude seed extracts of mung bean (Vigna radiata) contained a potent agglutinin of trypsinized[8] rabbit erythrocytes. These extracts showed no activity with normal or
trypsinized human (A, B or O) erthrocytes or with untryp-
sinized rabbit cells. They did exhibit two unique characteristics. First, hemagglutinin activity (MBH) was inhibited by xylose and myo-inositol, as well as galactose, a
specificity unlike that encountered in previously described
hemagglutinins. Secondly, these extracts possessed an interesting property, referred to here as "clot-dissolving"
activity.[4]

Since observation of clot-dissolving activity led
directly to the discovery of α-galactosidase-hemagglutinins
and also provided a diagnostic test for the presence of this
type of hemagglutinin, this phenomenon is described in some
detail. When a concentrated crude extract of mung bean
seeds was mixed with rabbit erythrocytes, the cells were
totally agglutinated within one minute, but within ten minutes the cells disaggregated and all evidence of hemagglutinin disappeared. Upon addition of a fresh sample of extract to the disaggregated cells they failed to re-aggregate.
If instead, fresh erythrocytes were added to the disaggregated cell mixture (or a dilution of same) agglutination
occurred. These observations suggested that the dissolution
of cell aggregates was due to an alteration of rabbit erythrocytes rather than an inactivation of the hemagglutinin.

When normal titer assays (serial two-fold dilutions)
were performed, the first dilution (most concentrated) gave
rise to cell aggregates which disappeared (disaggregated)

after about ten minutes. In the second serial dilution
agglutination persisted for about twenty minutes while in
the third dilution series aggregates were no longer evident
at the end of one hour. Agglutination was observed in all
subsequent dilutions to a final dilution of 1:2000 (titer =
2000). For twenty-four hours, the final titer remained
unchanged (2000), but cell aggregates in the fourth, fifth
and sixth serial dilutions had successively disappeared.
The final result of a twenty-four hour titer assay, then,
was agglutination in titer wells corresponding to dilutions
of about 1:128 to about 1:2000, but no agglutination at
dilutions of less than about 1:128.

The observations described above were made at room
temperature, but similar studies were done at 5°C and 37°C.
These studies revealed that the time required for a clot
(agglutinated cells) to disappear was extended at lower
temperatures (3 to 4 times longer at 5°C than at 22°C), but
considerably shortened at higher temperatures (about 1/2 as
long at 37°C as at 22°C).

The temperature and concentration dependence of clot-
dissolving activity suggested that it might be the result
of an enzymatic process. Therefore, the next step was to
examine mung bean extracts for glycosidase activities in an
attempt to determine if one or more of these enzymes were
responsible for clot dissolution. Glycosidases were assayed
using nitrophenylglycosides as substrates. Carbohydrates
were then tested for their ability to specifically inhibit
each activity that was found. If a specific glycosidase
were involved in clot dissolution, such inhibitors might be
used to block clot-dissolving activity during the hemagglutinin titer assay.

When the α-galactosidase activity in mung bean extracts
was examined it was found to be inhibited by galactose,
xylose and myo-inositol, a specificity identical to that
displayed by the hemagglutinin activity. Further, the substrate used for the assay of α-galactosidase, p-nitrophenyl
α-galactoside was a very potent inhibitor of the hemagglutinin activity. Other carbohydrates were tested but none
were found which inhibited clot-dissolving activity or
α-galactosidase activity without also inhibiting the hemagglutinin.

Since the hemagglutinin and α-galactosidase activities
displayed such similar specificities it appeared possible
that the enzyme might be recognizing and enzymatically
altering these cellular sites recognized by the hemagglutinin. Further studies along these lines, however, required
that the activities be purified.

Mung Bean α-Galactosidase-Hemagglutinin

Attempts to separate mung bean α-galactosidase from
hemagglutinin activity were unsuccessful. Both activities
exactly co-purified by every separation method tested. A
combination of ion exchange chromatography and gel filtration produced a preparation that contained a single protein
species with very high specific enzyme and hemagglutinin
activities. A summary of the overall purification is given
in Table 1.

The native protein had a molecular weight of about
160,000 and appeared to be composed of a single type of
subunit (45,000 mol. wt). The mung bean enzymatic hemagglutinin (MBH) was, therefore, assumed to be a tetramer of
identically sized subunits.[4]

Table 1. Summary of the purification of mung bean enzy-
matic-hemagglutinin

Step	Volume	Protein ?	Hemagglutinin A	α-Galactosidase B	Ratio A/B
	ml	mg/ml	HU/mg x 10^3	Units/mg	
Crude	2200	17.7	4.5	3.84	1.17
$(NH_4)SO_4$	300	57.0	6.3	5.15	1.03
CM pool	93	0.16	4,500	4,750	0.95
S-200 pool	18.5	0.13	18,600	14,400	1.36

Attempts to selectively inactivate one of the activities of the MBH by the use of heat or p-chloromercuri-
benzoate were unsuccessful. In fact, both activities were
equally sensitive to these treatments and equal protection
from inactivation was provided by galactose, xylose and
myo-inositol.[4]

MBH displayed linear Michaeles Menten kinetics with
galactose, xylose or myo-inositol acting as competitive
inhibitor of p-nitrophenyl α-galactoside utilization. A
summary of the kinetic parameters obtained is given in
Table 2. The carbohydrate specificity of the hemagglutinin
activity (Table 3) was virtually identical both qualita-
tively and quantitatively to that of the enzyme.

Attempts to separate the enzymatic and hemagglutinin
activities by adsorption on rabbit erythrocytes were unsuccessful.[4] Both of these activities were adsorbed by cells
under conditions favorable to hemagglutinin activity and
both were released from cells under conditions favoring
enzymatic activity. Further, both activities were released
from rabbit cells following the addition of galactose,
xylose or myo-inositol.

Recent studies revealed that MBH can be reversibly
converted by pH shifts, from a tetrameric, enzymatically
active hemagglutinin to an enzymatically active monomeric
protein that is devoid of hemagglutinin activity.[9] This
α-galactosidase therefore, also exists in two different
molecular weight forms as has been shown to be the case
with the enzymes from Vicia faba[10] and soybeans.[11] Although
both monomeric and tetrameric forms of MBH are enzymically
active, they display quite different pH optima, kinetic
properties and carbohydrate specificities.[9]

Soybean α-Galactosidase Hemagglutinin

Soybean (Glycine max.) seeds and leaves were found to
contain a clot-dissolving hemagglutinin activity.[12] In
seeds, this activity cannot be visualized without first inhibiting or removing (by affinity chromatography) SBA, the
well characterized N-acetylgalactosamine specific hemagglutinin of soybeans. The soybean clot-dissolving hemagglutinin has been purified to homogeneity by both conven-
tional procedures and by affinity chromatography on melibiose

Table 2. Kinetic parameters of the α-galactosidases in
        several legumes[1]

Non-Hemagglutinin Forms	Km p-Nitrophenol α-Gal mM	Ki		
		Galactose	Xylose mM	Inositol
Amorpha	1.25	0.53	2.95	12.6
Bandeiraea s.	1.11	0.62	4.45	14.7
Bauhinia p.a.	1.00	0.50	3.10	15.8
Caragana a.	0.80	0.61	3.85	9.5
Cercis s.	1.82	0.47	3.40	9.0
Colutea a.	0.44	0.37	3.20	17.3
Conavalia e.	1.18	0.22	1.75	9.5
Cytisus m.	1.00	0.30	1.85	6.1
Dolichos b.	1.18	0.51	3.60	--
Genista m.	0.58	0.35	1.60	11.0
Laburnum a.	1.43	0.53	2.75	14.7
Lathyrus l.	1.33	0.61	3.50	8.7
Lens c.	1.21	1.19	3.30	6.9
Lespedeza b.	0.47	0.35	2.70	11.7
Mimosa p.	0.65	0.38	2.2	11.6
Phaseolus v.	0.71	0.50	3.15	12.8
Sophora j.	0.57	0.39	2.25	9.0
Spartium j.	1.66	0.45	3.55	11.4
Ulex e.	0.67	0.37	1.75	10.0
Wistaria s.	0.65	0.58	3.65	17.8
Hemagglutinin forms				
Mung bean	0.20	0.75	5.2	20.0
Pueraria	0.42	0.90	2.9	8.8
Thermopsis	0.20	0.40	1.6	8.8
Lupine	0.39	0.30	1.5	7.5
Lima bean	0.38	0.33	2.0	15.0
Soybean	0.33	1.06	3.80	12.9

[1]Data collected from references 4, 9, 12 and 13.

substituted sepharose.[13] As with the mung bean protein,
this hemagglutinin was found to exactly co-purify with an
α-galactosidase activity.

Table 3. Inhibitor specificities of the hemagglutinin activities of several α-galactosidase-hemagglutinins

Inhibitor	Mung beans	Soybean	Pueraria	Thermopsis	Lupine	Lima beans
			mM^a			
p-Nitrophenyl α-galactosidase	0.34	0.1	1.0	1.0	3.1	0.6
Galactose	1.0	0.38	1.5	0.8	3.1	0.8
Xylose	4.0	0.76	3.0	3.0	6.3	6.2
Inositol	20.0	3.06	25.0	25	25	31
Galactosamine	> 50	> 50	> 50	> 50	> 50	> 50

^{a}Minimum concentration required to totally inhibit four hemagglutinin units.

Detailed characterization of the soybean α-galactosidase hemagglutinin revealed that it was nearly identical to the mung bean protein in most respects (see data in Tables 2, 3 and 4) and was immunologically indistinguishable from same.[13] One major difference between the two proteins was noted; the soybean α-galactosidase was composed of two very similar but distinct subunits with molecular weights of about 38,000 and 40,000. The soybean protein displayed pH dependent association-disassociation behavior similar to that described for mung bean, and possesses the same aggregation dependent variations in enzymatic and hemagglutinin properties.

Soybean α-galactosidase hemagglutinin is probably the same α-galactosidase purified and characterized by Harpaz et al.[11] however, these authors did not describe any associated hemagglutinin activity, nor did they discuss the changes in enzymic properties that accompany changes in aggregation state.

Table 4.   Summary of gel filtration molecular weight deter-
minations for various legume α-galactosidases

Plant	α-Galactosidase I	II[1]
Non-hemagglutinin forms		
Colutea	135,000	34,000
Cytisus	190,000	40,000
Genista	160,000	37,000
Laburnum	155,000	30,000
Lathyrus	160,000	--[2]
Lespedeza	120,000	31,000
Sophora	150,000	--[2]
Spartium	160,000	30,000
Ulex	160,000	42,000
Hemagglutinin forms		
Mung bean	160,000	45,000[3]
Pueraria	150,000	--[2]
Thermopsis	180,000	--[2]
Lupine	190,000	50,000
Lima bean	180,000	39,000
Soybean	160,000	38,000 and 40,000[3]

[1] Galactosidase II may generally be representative of a
monomeric (subunit) form of α-galactosidase I.   This
appears to be the case in mung beans and soybeans.
[2] α-Galactosidase II not observed in crude extracts of these
species.
[3] Determined by SDS gel electrophoresis of pure α-galactosidase I.

Other α-Galactosidase-Hemagglutinins

In a survey of 27 legume species from 26 genera four plants in addition to mung bean and soybean were found to contain clot-dissolving hemagglutinins (Table 5).[12] These clot-dissolving hemagglutinins exactly copurified by ion exchange and gel filtration chromatography with an α-galactosidase activity.[12] From the data in Tables 2, 3, and 4, it can be seen that the physical, enzymatic and hemagglutinin properties of these proteins are very nearly identical to each other and to MBH.

The α-galactosidase hemagglutinin in lima bean is a distinctly different protein from the well characterized N-acetylgalatosamine specific hemagglutinin that is present in lima bean.[12] We have examined three lima bean varieties (Burpee "Bush", "Fordhook", and "Sieva") and found the clot-dissolving, enzymic hemagglutinin in all three. The N-acetylgalactosamine specific lima bean hemagglutinin however, appears to be present in significant quantities only in the "Sieva" variety of lima bean.

Non-Agglutinating α-Galactosidases

From Table 5 it can be seen that most legume species examined did not contain clot-dissolving hemagglutinin activity. In fact many plants appeared to be devoid of hemagglutinin activity altogether.[14] All species examined, however, contained α-galactosidase activity. The kinetic and physical properties of the non-agglutinating α-galactosidases were examined to determine if they were related to the hemagglutinating forms.

Kinetic studies (summarized in Table 2) showed that the enzyme in each extract displayed simple Michaelis Menton substrate utilization with competitive inhibition by galactose, xylose and myo-inositol.[14] The striking feature of this data is the remarkable similarities, not only among the non-agglutinating α-galactosidases, but also between these forms and the α-galactosidase-hemagglutinins.

Gel filtration experiments revealed that all plants examined contained a large molecular weight enzyme activity and many also contained a small molecular weight form. From the data in Table 4 one can readily see the similarities in

Table 5. α-Galactosidase and related activities in several legume species

Plant	α-Galactosidase units/g dry seed	Hemagglutinin activity of any kind	Clot-dissolving hemagglutinin activity	MBH-CRM
Amorpha fruticosa	22	No[1]	No	No[2]
Cercis siliquastrum	39	No	No	Yes
Colutea arborescens	104	No	No	Yes
Genista monosperma	450	No	No	Yes
Lespedeza bicolor	300	No	No	No
Mimosa pudica	75	No	No	Yes
Spartium junceum	50	No	No	Yes
Gymnocladus	4	No	No	No
Vigna radiata	111	Yes	Yes	Yes
Glycine max	62	Yes	Yes	Yes
Pueraria thunbergiana	110	Yes	Yes	Yes
Thermopsis caroliniana	116	Yes	Yes	Yes
Lupinus arboreus	43	Yes	Yes	Yes
Phaseolus limensis	26	Yes	Yes	Yes
Bandeiraea simplicifolia	20	Yes	No	No
Bauhinia purpurea alba	72	Yes	No	Yes
Caragana arborescens	200	Yes	No	Yes
Conavalia ensiformis	225	Yes	No	No
Cytisus multiflorus	36	Yes	No	Yes
Dolichos biflorus	60	Yes	No	Yes
Lathyrus latifolia	300	Yes	No	Yes
Lens culinaris	175	Yes	No	No
Phaseolus vulgaris	60	Yes	No	Yes
Sophora japonica	140	Yes	No	Yes
Wistaria sinensis	180	Yes	No	Yes

[1] As determined by the visual agglutination of trypsinized rabbit or human (A, B and O) erythrocytes.

[2] As determined by Ouchterlony double diffusion of extracts with antisera raised against pure Vigna α-galactosidase-hemagglutinin.

molecular weights, no only among the non-hemagglutinin forms,
but between these enzymes and the α-galactosidase hemagglutinins.

Immunochemical Studies

Immunological evidence has been presented which re-
veals that the various α-galactosidase-hemagglutinins are
evolutionarily closely related.[12] In fact, those from mung
bean and soybean appear to be identical immunologically.[13]
Similarities in physical and kinectic properties among these
proteins as well as their related immunological properties
suggest a strong homology.

A survey of several legume species (Table 5) revealed
that seed extracts of most plants tested contained cross-
reacting materials (CRM) which were immunologically related
to MBH (MBH-CRM), even though most of the plants did not
contain α-galactosidase hemagglutinins. Immunoaffinity
chromatography was used to determine if the non-agglutinating α-galactosidases were responsible for the MBH-CRM
seen in these plants. In this method IgG was purified from
both anti-MBH sera and pre-immune sera. Each IgG was then
coupled to sepharose 4B. Crude extracts from different
legume species, all containing α-galactosidase activity,
were passed over columns of pre-immune IgG sepharose and
anti-MBH IgG sepharose. The results (Table 6) show that
α-galactosidase activity is specifically absorbed from every
extract by anti MBH IgG but not by pre-immune IgG sepharose.
No specific binding was seen for several other glycosidases
tested. By using a larger quantity of column adsorbant or
by prolonged incubation of the extracts with adsorbant,
virtually 100% of the α-galactosidase activity could be
adsorbed from each of the extracts. These studies[14] clearly
indicate that non-agglutinating α-galactosidases are closely
related immunologically to the α-galactosidase hemagglutinins. This result coupled with the great similarities in
their physical and kinetic properties suggests that both
types of α-galactosidases are members of one specific
functional class of protein (i.e., they are homologues).

The evidence collected to date indicates that the
α-galactosidase hemagglutinins and the "classic" legume
hemagglutinins are members of distinct classes of proteins.[12]
Immunological studies[15] reveal, however, that these two

Table 6. The binding of α-galactosidase activities in crude
legume extracts to anti <u>Vigna</u> IgG-Sepharose

Plant	Protein	α-Galactosidase	β-Galactosidase	β-Glucosidase	α-Mannosiadse
		% Specific Binding*			
1. Sophora	0	95	2	0	0
2. Lespedeza	0	70	9	0	0
3. Amorpha	0	68	0	0	0
4. Caragana	0	66	0	0	0
5. Ulex	0	71	0	0	0
6. Colulea	0	60	0	0	0
7. Lens	0	80	0	0	0
8. Conavalia	0	40	0	0	0

* Computed as the percentage of activity adsorbed to the
specific IgG column minus the percentage bound to the
preimmune IgG column.

classes of proteins are evolutionarily related (i.e., share
at least some antigenic determinants). This fact derives
from the observation that antisera to several "classic"
lectins cross react with MBH and antiserum to MBH cross
reacts with several classic lectins. It should be noted
that relatively high concentrations of antigen and strong
antisera are required in order to see these cross reactions.

α-MANNOSIDASE-HEMAGGLUTININS

Paus and Steen[5] have described an α-mannosidase from Phaseolus vulgaris which displayed hemagglutinin properties. The hemagglutinin has a molecular weight of about 220,000 and apparently is a glycoprotein composed of two non-covalently bound subunits of 110,000 molecular weight. This protein has been studied in some detail[16] [17] with respect to its enzymatic and physical properties, but very little has been reported about its hemagglutinin properties. Likewise its relationship (if any) to the classic hemag-glutinin (PHA) in P. vulgaris or to other legume hemag-glutinins is not known. The authors did, however, show that P. vulgaris α-mannosidase was a mitogen with a potency com-parable to other legume mitogenic hemagglutinins.[5] Evidence was also provided suggesting that similar enzymic-mitogenic hemagglutinins may exist in mammalian tissues.

N-ACETYL GALACTOSAMINE-SPECIFIC HEMAGGLUTININS

Immunological Surveys

A great many of the legume hemagglutinins that have been studied show specificites for N-acetyl-galactosamine (GalNAc lectins).[3] Among those which have been character-ized are the GalNAc lectins from species of Bauhinia, Dolichos, Glycine (soybean), Phaseolus (lima bean), Wisteria, Sophora and Caragana. The GalNAc lectins from most or all of these legume species are glycoproteins, bind metal ions, exist in tetrameric forms, are isolated from seeds and have similar sized subunits.[17] The great overall similarities in these proteins suggests that they may be homologues. This possibilty is very strongly supported by the finding that these proteins are immunologically related[15,19] and may generally possess a high degree of amino acid sequence homology.[20,21] We have recently completed N-terminal amino acid sequence studies with GalNAc lectins from Bauhinia, Caragana, Sophora and the fucose specific lectin from Ulex and found that these proteins all show extensive sequence homology with one another and with previously analyzed lectins.[20,21]

In view of the above findings and the fact that GalNAc lectins often account for 1% or more of the seed protein, and are widely distributed among legume species,[2] we

wondered if they might be ubiquitious legume proteins. A
review of the literature suggested that this possibility was
unlikely, since many legume species (perhaps 50% or more of
all species) appeared to be totally devoid of hemagglutinin
activity of any kind. Furthermore, many of the species which
contained hemagglutinins appeared to contain proteins with
carbohydrate specificities distinct from the GalNAc lectins.[2]
To confirm the validity of the earlier studies a survey of a
number of legume species was performed and, indeed, seed
extracts of many legume species displayed very little or no
hemagglutinin activity (Table 7) or contained hemagglutinins
which werenot inhibited by GalNAc (unpublished data). One
cannot rule out the presence of a hemagglutinin activity by
the limited assay methods used. However, since all GalNAc
hemagglutinins which have been described are very easily
detected in crude extracts by the assay methods used, it is
reasonable to conclude that none of these plants contain a
detectable quantity of hemagglutinin activity comparable to
the GalNAc hemagglutinins.

Table 7. Hemagglutinin content of crude legume seed extracts

| | Hemagglutinin Titers Blood Type[1] | | | |
| | Rabbit | Human | | |
Plant Species	R	A	B	O
1. Acacia decurrens mollis	0	0	0	0
2. Amorpha fruticosa	0	0	0	0
3. Cercis siliquastrum	0	0	0	0
4. Colutea arborescens	0	2	2	8
5. Delonix regia	8	4	2	2
6. Genista monosperma	0	0	0	0
7. Gleditsia triacanthos	4	4	2	2
8. Lupinus polyphyllos	8000	0	0	0
9. Lespedeza bicolor	0	0	0	0
10. Mimosa pudica	4	0	2	0
11. Parkinsonia aculeata	16	4	4	4
12. Poinciana gilliesi	2	0	0	0
13. Pueraria thunbergiana	2000	0	0	0
14. Spartium junceum	0	0	0	8
15. Thermopsis caroliniana	500	0	0	0

1. All assays were done using trypsinized[8] erythrocytes.

We decided to examine extracts of a number of GalNAc hemagglutinin "negative" species to determine if they contained proteins (CRM) immunologically related to any of the GalNAc hemagglutinins. Results (Table 8) reveal that every plant tested contained CRM to one or more of the GalNAc hemagglutinins tested. The antisera used in these studies were from rabbits immunized with very pure hemagglutinin preparations. Although many plants showed CRM with antisera raised against several different lectins, it appeared likely that one major protein was responsible since adsorption of the extracts by any one of the reactive sera results in a loss of reactivity with all the other sera (unpublished observations).

Since the GalNAc hemagglutinins are glycoproteins, the antisera raised against them might recognize glycoproteins in plant extracts which are totally different from the hemagglutinins, but which possess similar carbohydrate determinants. This possibility was ruled out in the following way: The quantity of hemagglutinin (antigen) required to totally inhibit the homologous antisera was determined. About 100 times this quantity of antigen was denatured by boiling and then subjected to several proteolytic treatments. The resulting material retained no hemagglutinin activity, showed no immunological reaction by Ouchterlony double diffusion and most importantly, showed no ability to block the reaction of antisera with CRM containing extracts. All the buffers and protease solutions were tested for the presence of several glycosidases and none were found. If carbohydrate determinants constituted a dominant part of the CRM reaction, then the glycopeptides remaining in the proteolyzed antigen preparation would be expected to have potent hapten inhibitors of the CRM reaction. Since no inhibition was seen, it was concluded that carbohydrate plays little or no role in determining the immunological similarities between hemagglutinins and their CRMs (unpublished observations).

Affinity Chromatographic Studies

Most plant hemagglutinins can be purified (or partially purified) by affinity chromatography.[13] Sepharose to which N-acetyl galactosamine has been attached provides an excellent adsorbant for the purification of most of the GalNAc specific hemagglutinins. There are now many examples of proteins which have carbohydrate binding properties, but

Table 8.   Immunological cross reactions between crude seed
extracts and antisera raised against purified
GalNAc specific legume lectins

Seed Extract	Antisera*		
	Bauhinia	Dolichos	Sophora
1. Acacia	+	+	+
2. Amorpha	+	−	+
3. Cercis	±	−	+
4. Colutea	+	−	+
5. Delonix	+	±	+
6. Genista	+	−	+
7. Gleditsia	+	−	+
8. Lupinus	+	−	+
9. Lespedeza	+	−	+
10. Mimosa	+	+	+
11. Parkinsonia	±	−	+
12. Poinciana	+	−	+
13. Pueraria	+	+	+
14. Spartium	+	−	+
15. Thermopsis	+	−	+
Total positive CRM	13	3	15

* + = precipitin line observed

  − = no precipitin line observed

  ± = very weak reaction, questionable

which are not themselves hemagglutinins, that can be purified by carbohydrate affinity chromatography. It is also
known that hemagglutinins may sometimes exist in forms (or
can be converted to forms) which retain carbohydrate binding
properties but are no longer hemagglutinins. This information leads one to question whether or not the CRMs, even
though they do not display hemagglutinin activity, can be
adsorbed by affinity matricies. Therefore, a number of CRM
containing plant extracts were chromatographed on GalNAc
sepharose columns with the result that very little, if any
of the total protein and no CRM was retained by the columns
(unpublished observations). These findings suggest that
the CRMs, which do not display classic hemagglutinin activity,
also do not possess the carbohydrate binding properties
typical of GalNAc lectins.

Immunoadsorption Studies

The results obtained using the immunoadsorption methods
developed for the study of α-galactosidases suggested that
similar techniques might be useful to further characterize
GalNAc lectin CRMs. We therefore prepared IgG-sepharose
columns using purified IgG from pre-immune sera and from
several anti GalNAc lectin sera.

Columns containing pre-immune IgG did not bind CRMs
(i.e., adsorb it from crude extracts). As expected, however,
CRMs were very effectively adsorbed by columns containing
IgG from cross reactive antilectin sera (unpublished observations). We could now determine whether any specific proteins (such as enzymes) were selectively adsorbed from
extracts by anti-lectin IgC columns. Most of the work was
done using IgG purified from antisera against the GalNAc
lectins from Bauhinia and Sophora.

Very little of the total protein in extracts was adsorbed by either pre-immune or antilectin IgG sepharose.
However, when glycosidase activities were followed, it appeared that in extracts from several legume species a
β-hexosaminidase activity was specifically adsorbed by the
antilectin IgG columns (unpublished observations). This
very interesting result led us to ask if plants containing
GalNAc lectins, such as Sophora and Bauhinia also contained
β-hexosaminidases and where these enzymes were related to
the lectins.

Our studies in this area are just beginning so to date are preliminary. Despite unsuccessful attempts to purify the β-hexosaminidases from Sophora japonica and Bauhinia purpurea alba, some observations were made as regards to these activities:

1. The enzymes from each plant have molecular weights (by gel filtration on Sephacryl S-200) which are practically identical to those of the respective lectins.

2. The enzyme and lectin in each plant have very similar (but not identical) ion exchange behavior.

3. The lectin is adsorbed from extracts by the affinity adsorbent, GalNAc Sepharose, but not the β-hexosaminidases.

4. Material not retained by GalNAc sepharose, i.e., the lectin-depleted fraction still contains material immunologically related to the lectin.

5. β-Hexosaminidase activity in Sophora extracts from which the lectin has been removed by affinity chromatography, adsorbs to anti Sophora IgG columns but not to pre-immune IgG sepharose. The same results are seen with the Bauhinia enzyme.

Our studies suggest the legume β-hexosaminidases and GalNAc specific lectins are evolutionarily related proteins. If these observations are substantiated by further investigation, then studies of the relationship between these two GalNAc binding proteins should provide a new and rewarding approach toward understanding legume hemagglutinins.

CONCLUDING REMARKS

The evidence at this time suggests that there are at least two evolutionarily related but functionally distinct classes of legume protein which possess hemagglutinin properties. These are the "classic" lectins and the α-galactosidase hemagglutinins. Whether or not the α-mannosidase hemagglutinin seen in Phaseolus constitutes a member of a third class remains to be seen. Also it is not yet clear whether the "classic" lectins are all homologues or proteins from several distinct functional classes. However, the GalNAc specific lectins that have been described certainly appear to be homologues. This is an important conclusion because we can now refer to these proteins (GalNAc lectins) as members of a specific physiologically functional class even though we do not know their function.

In comparing the studies of the GalNAc lectins with findings obtained for the α-galactosidase-hemagglutinins, a number of analogies are readily apparent. We know that all legumes contain a specific α-galactosidase but that only rarely does this enzyme possess hemagglutinin activity. Likewise, it appears that all legumes contain a GalNAc-lectin-like protein (CRM) but only occasionaly contain a GalNAc specific hemagglutinin activity. We suggest that all legumes probably contain a functional homologue of the GalNAc lectins and that it is this protein which gives rise to the majority of the GalNAc lectin CRM seen in many legumes. Thus, just as there may be hemagglutinin "active" and "inactive" forms of α-galactosidase, there may be hemagglutinin "active" and "inactive" forms of the GalNAc lectins. We further suggest that the "hemagglutinin activity" of the GalNAc lectins, just as with the α-galactosidase hemagglutinins, is probably not important with respect to the general functioning of this class of proteins.

Our studies with β-hexosaminidases strongly suggest that this enzyme is closely related to the GalNAc lectins. Just how these two kinds of GalNAc binding proteins are related remains to be seen. They might be functionally distinct but evolutionarily related proteins, as appears to be the case between the α-galactosidases and the classic lectins or they could be more closely related, perhaps displaying a precursor-product relationship.

Independent of the ultimate validity of the foregoing speculations, we feel the current body of knowledge indicates that great care must be exercised with respect to the terms lectins and hemagglutinin. A variety of proposals have been advanced with regard to the function of plant lectins,[7] however, virtually all these theories rely on the assumption that hemagglutinin activity is a direct reflection of the in vivo activity of these proteins. This assumption is not necessarily valid and by discarding it, one can view lectins from a much broader perspective. For example, one might view lectins as functionally inactive or altered forms of specific plant proteins, in which case hemagglutinin activity might represent little more than an artifactual property. The idea that hemagglutinin activity may not be a required manifestation of the physiological funtion (activity) of these proteins is not unreasonable since their in vivo role is totally unknown and thus no specific assay for this function exists. One cannot safely define a class of proteins with respect to a physiological function by a criterion (hemagglutinin activity) which has never been shown to be a direct reflection of that function. Thus, one should not conclude that a plant is devoid of a physiologically active "lectin" simply because it may lack a protein with hemagglutinin activity or may contain a protein with activities in addition to hemagglutinin activity.

The rapidly expanding area concerning immunological and amino acid sequence homologies is a good trend, we think, and may offer the best prospects for overcoming the difficulties associated with terminology. This is particularly relevant here because these methods are "blind" with respect to functions or activites and yet, can reveal much about the "relatedness" of proteins. Clearly a great many questions about these interesting and yet elusive proteins remain to be answered. We hope our ideas and observations will stimulate additional work from broader perspectives, which will provide the necessary foundation from which their physiological roles will ultimately emerge.

ACKNOWLEDGEMENTS

We expresss sincere gratitide to Juanita I. Kindinger, Kevin Mielke and Theresa Canfield for their excellent technical assistance during various phases of this work. We also wish to indicate that Elena Del Campillo, during the course of her thesis research made significant contributions in all areas of our studies, in addition to those which are specifically cited.

REFERENCES

1. Stillmark. 1888. Uber Rizin, ein giftiges ferment aus dem Samen von Ricinus communis L. und einigen aderen Euphorbiaceen. Inaug. Diss. Dorpat.

2. Toms, G. C. 1971. Phytohemagglutinins. In:
    Chemotaxonomy of the leguminosase. (J. B. Harbourne,
    D. Boulter, B. L. Turner, eds.) Academic Press, New
    York. pp. 367-462.
3. Goldstein, I. J. and C. E. Hayes. 1978. The lectins:
    carbohydrate-binding proteins of plants and animals.
    Adv. Carbohyd. Chem. Biochem. 35:127-340.
4. Hankins, C. N. and L. M. Shannon. 1978. The physical
    and enzymatic properties of a phytohemagglutinin from
    mung beans. J. Biol. Chem. 253:7791-7797.
5. Paus, E. and H. B. Steen. 1978. Mitogenic effect of
    α-mannosidase of lymphocytes. Nature 272:452-454.
6. Lis, H. and N. Sharon. 1973. The biochemistry of
    plant lectins (Phytohemagglutinins). Annu. Rev.
    Biochem. 42:541-574.
7. Liener, I. E. 1976. Phytohemagglutinins (Phytolectins).
    Annu. Rev. Plant Physiol. 27:291-319.
8. Lis, H. and N. Sharon. 1972. Soy bean (Glycine max)
    agglutinin. Methods Enzymol. 28B:360-365.
9. Del Campillo, E. and L. M. Shannon. 1980. Aggregation
    dependent properties of the enzymic lectin from mung
    beans. Pacific Slope Biochemical Conference, Univ. of
    Calif., San Diego. 1980:19.
10. Dey, P. M. and J. B. Pridham. 1969. Purification and
    properties of α-galactosidases from Vicia faba seeds.
    Biochem. J. 113:49-55.
11. Harpaz, N., H. M. Flowers and N. Sharon. 1977.
    α-D-galactosidase from soybeans destroying blood-group
    B antigens. Eur. J. Biochem. 77:419-426.
12. Hankins, C. N., J. I. Kindinger and L. M. Shannon.
    1980. Legume α-galactosidases which have hemagglutinin
    properties. Plant Physiol. 65:618-622.
13. Del Campillo, E. 1980. Legume Hemagglutinins with
    α-Galactosidase Activity, Ph.D. Thesis, Department of
    Biochemistry, Univ. of Calif., Riverside.
14. Hankins, C. N., J. I. Kindinger and L. M. Shannon.
    1980. Legume α-galactosidase forms devoid of hemag-
    glutinin activity. Plant Physiol. 66:375-378.
15. Hankins, C. N., J. I. Kindinger and L. M. Shannon.
    1979. Legume lectins I. Immunological cross-reactions
    between the enzymic lectin from mung beans and other
    well characterized legume lectins. Plant Physiol.
    64:104-107.

16. Paus, E. 1976. Immunoadsorbent affinity purifica-
     tion of the two enzyme forms of α-mannosidase from
     Phaseolus vulgaris. FEBS Lett. 72:39-42.
17. Paus, E. 1977. α-Mannosidase from Phaseolus vulgaris:
     Composition and structural properties. Eur. J. Biochem.
     73:155-161.
18. Galbraith, W. and I. J. Goldstein. 1972. Phytohemag-
     glutinin of the lima bean (Phaseolus lunatus). Iso-
     lation, characterization and interaction with type A
     block group substance. Biochemistry 11I:3975-3984.
19. Howard, J., J. I. Kindinger and L. M. Shannon. 1979.
     Conservation of antigenic determinatnts among different
     seed lectins. Arch. Biochem. Biophys. 192:457-464.
20. Foriers, A., C. Wuilmart, N. Sharon and A. D. Strasberg.
     1977. Extensive sequence homologies among lectins from
     Leguminous plants. Biochem. Biophys. Res. Commun.
     75:980-986.
21. Etzler, M., C. F. Talbot and P. R. Ziaya. 1977.
     NH$_2$-Terminal sequences of the subunits of Doliches
     Biflorus lectin. FEBS Lett. 82:39-41.

Chapter Six

RECOVERY OF GLYCOPROTEINS FROM PLANT TISSUES

ROBERT G. BROWN AND W. C. KIMMINS

Department of Biology
Dalhousie University
Halifax, Nova Scotia, Canada

INTRODUCTION

The study of plant glycoproteins received a major stimulus by the discovery of plant lectins and their commercial
exploitation. Many lectins are glycoproteins and some have
structural similarities with cell wall glycoproteins. Some
glycoproteins have well-defined and important biological
functions, whereas, the function of other glycoproteins
remains elusive. This is particularly true of cell wall
glycoproteins which are widely distributed among plants suggesting an important function which has yet to be discovered.

Recovery of glycoproteins from plant tissue poses some
unique problems, particularly, if the glycoproteins are to
be obtained intact. This is particularly true of cell wall
glycoproteins which are firmly bound to the wall and generally must be degraded before extraction is possible. Cell
walls of higher plants contain at least two glycoproteins,
termed hydroxyproline-rich glycoprotein[1] or extensin[2] and
hydroxyproline-poor glycoprotein.[1] The main glycoprotein
of Chlamydomonas reinhardii may be cleaved into hydroxyproline-rich and hydroxyproline-poor fragments suggesting that
hydroxyproline-rich and hydroxyproline-poor glycoprotein may
be parts of a larger complex.[3] Wounding leaves of Phaseolus
vulgaris increases production of a hydroxyproline-poor

115

glycoprotein which binds to cellulose and therefore may have
a cross-linking role in the cell wall.[4-8] Methods of ex-
tracting hydroxyproline-rich and hydroxyproline-poor glycoprotein will be examined in this article as a means of
assessing techniques available for the extraction of glycoproteins involved in recognition phenomena which are associated with the cell wall. Although recovery of both hydroxyproline-rich and hydroxyproline-poor glycoproteins will be
discussed, emphasis will be placed on the latter glycoprotein. In particular, the distribution of hydroxyproline-
poor glycoprotein among land plants and within plant cells
will be emphasized. An attempt will be made to explore the
relationship between hydroxyproline-poor glycoprotein found
in the cell wall and that produced in response to wounding.

EXTRACTION AND PROPERTIES OF GLYCOPROTEINS

     Although many methods have been used to extract glyco-
proteins from plant tissue, not all are suitable for extract-
ing hydroxyproline-rich or hydroxyproline-poor glycoprotein.
Hydroxyproline-rich glycoprotein has been extracted from
hypocotyl cell walls with alkali[9,10] and cell walls from
other tissue with sodium chlorite.[1,11,12] In addition,
sodium perchlorate has been used to extract hydroxyproline-
rich glycoprotein from algal cell walls.[3] Hydroxyproline-
rich glycoprotein extracted with sodium chlorite from cell
walls of P. vulgaris has structural features summarized in
Fig. 1. The glycoprotein is low molecular weight with
hydroxyproline as the most abundant amino acid, twice as
much arabinose as galactose and the glycosidic linkages
expected for a tetraarabinoside-containing structure. The
presence of (1→4)-linked glucose residues is unexpected but
this has also been found in a similar preparation from P.
coccineus.[11] Hydroxyproline-poor glycoprotein has been extracted from wounded leaf tissue with alkali[4] and from cell
walls with sodium chlorite.[1,12] Following extraction,
hydroxyproline-poor glycoprotein may be purified by gel
filtration and isoelectric focusing (Fig. 2).[4,13] Hydroxyproline-poor glycoprotein from P. vulgaris has the structural
features summarized in Fig. 3. This glycoprotein is high
molecular weight, with lysine as the only N-terminal amino
acid suggesting the presence of one polypeptide chain.
Arabinofuranose residues are linked (1→2), (1→3) and (1→5),
galactose is terminal and glucose occurs as (1→4)-linked
units. Hydroxyproline-poor glycoprotein binds to cellulose,

Amino acids (Mol. %)			Carbohydrate (Mol. %)	
Hyp	15		Ara	50
Asx	12		Xyl	3
Thr	11		Man	5
Ser	12	**Protein-Carbohydrate linkage**	Gal	27
Glx	6		Glc	15
Pro	5	Ser - gal (α-linked)		
Gly	10	Hyp - ara, gal (ara, β-linked)	**Linkage (Mol. %)**	
Ala	5			
Val	4		Ara (terminal)	16
Met	-		Ara (1 → 2)	24
Ilc	2		Ara (1 → 3)	12
Leu	2	**Molecular Weight**	Ara (1 → 5)	2
Tyr	1		Xyl (terminal)	3
Phe	1	17,000	Gal (terminal)	5
Tryp	3		Gal (1 → 4)	7
Lys (lys + AAA)	9		Glc (terminal)	2
His	1		Glc (1 → 4)	27
Arg	2			
Cys (Cys A)	1			

Figure 1. Structural features of hydroxyproline-rich glycoprotein from Phaseolus vulgaris (extracted with
sodium chlorite[1]).

Leaves
⃒ (maceration, Waring blender (5 min), Polyton (2 min), centrifuge).
Residue
⃒ (wash 3 x's cold water, 2°C).
Residue
⃒ (water, 100°C, 30 min).
Residue
⃒ (1 M NaOH, 2°C, 30 min).
Supernatant
⃒ (neutralize with acetic acid, dialyse).
Precipitate
⃒ (dissolve in 10% pyridine in water).
Sephadex G-150 (0. 02M phosphate buffer, pH 8.5).
⃒ (excluded fraction, freeze dry, rechromatograph).
Sephadex G-200 (0. 02M phosphate buffer, pH 8.5).
⃒ (excluded fraction dialyse, use as light solution for
⃒ isoelectric focusing without drying).
Isoelectric focusing
⃒ (wash or dialyse, depending on solubility of glycoprotein bands).

Hydroxyproline-poor glycoprotein

Figure 2. Extraction and purification of hydroxyproline-
poor glycoprotein.

(Matteucia strupthiopteris) dextran. The latter may be con-
veniently purified using Sephadex G-150. The glycoprotein

agarose, and, in one case binds to the gel during chromatography; the gel containing the bound glycoprotein is removed
from the column and eluted with 1% α-methylglucoside or
β-methylglucoside to yield the glycoprotein.

Most hydroxyproline-poor glycoproteins purified by isoelectric focusing were enriched with glucose following this
purification procedure (Table 1). Loss of xylose and mannose
following isoelectric focusing indicates that these sugars
were probably impurities. The isoelectric points of hydroxyproline-poor glycoproteins from four sources were similar;
all were acidic proteins. Isoelectric focusing (pH 3-10)
of glycoproteins from M. strupthiopteris, Rheum rhaponticum
and P. vulgaris was terminated after 18 hours, because
focusing for longer times resulted in precipitation of glycoproteins which settled to the bottom of the column. However,
the glycoprotein from Nicotiana tabacum did not precipitate
and it was possible to focus until equilibrium was achieved
(48 h, pH 2.5 - 5.0). Additional unidentified compounds detected during carbohydrate analyses following isoelectric
focusing indicated that ampholytes may bind to the glycoproteins. In addition, substantial increases in the apparent
lysine and histidine content of glycoproteins form M.

Amino acids (Mol. %)

			Carbohydrate (Mol. %)	
			Ara	28
Hyp	2		Xyl	5
Asx	7	**Protein-Carbohydrate linkage**	Gal	36
Thr	5		Glc	31
Ser	7	Ser - Ara, Gal, Glc		
Glx	10	Hyp - Ara, Gal, Glc	**Linkage (Mol. %)**	
Pro	7		Ara (terminal)	4
Gly	11		Ara (1 → 2)	9
Ala	11		Ara (1 → 3)	10
Val	6	**Molecular weight**	Ara (1 → 5)	9
Met	1	520,000	Xyl (terminal)	
Ilc			Xyl (1 → 4)	5
Leu	7	**Isoelectric point** 3.6	Gal (terminal)	18
Tyr	4		Glc (terminal	2
Phe	4	**N-terminus** lysine	Glc (1 → 4)	41
Tryp	1			
Lys	7			
His	1			
Arg	4	**Binding:** Cellulose, Agarose; Dextran **(M. struthiopteris)**		

Figure 3. Structural features of hydroxyproline-poor glycoprotein from Phaseolus vulgaris (extracted with
alkali[4]).

Table 1.   Sugar composition of hydroxyproline-poor glycoproteins before and after isoelectric focusing.

	Source								
Sugars	M. strupthiopteris		N. tabacum		R. rhaponticum			P. vulgaris	
%	Before	pI 3.2	Before	pI 3.7	Before	pI 4.5	pI 5.0	Before	pI 3.6
Ara	4	12	-	-	26	21	10	11	6
Xyl	2	-	4	-	-	-	-	13	-
Man	4	-	15	-	-	-	-	-	-
Gal	6	-	-	-	40	49	62	21	6
Glc	81	88	81	100	30	30	28	55	88

pI; isoelectric point.

strupthiopteris and R. rhaponticum and glycine in the glycoprotein from N. tabacum were detected following isoelectric
focusing. For instance, the glycine content of N. tabacum
glycoprotein was 57% after isoelectric focusing. Therefore,
isoelectric focusing was employed, in conjunction with gel
electrophoresis, to verify homogeneity, rather than as a
preparative procedure.

OCCURRENCE OF GLYCOPROTEINS

Hydroxyproline-arabinosides are present in cell walls
of plants from algae to angiosperms suggesting a wide distribution of hydroxyproline-rich glycoprotein among plants.[14]
Likewise, a study of the distribution of hydroxyproline-poor
glycoprotein has suggested a wide distributon of this glycoprotein among land plants.[13] Representatives of Pteridophyta,
Gymnospermae and Angiospermae contain hydroxyproline-poor
glycoproteins which have similar amino acid compositions
(Table 2). Linkage analysis indicates that the glyco-
moieties of glycoproteins are also similar in structure
(Table 3). In addition, trifluoroacetolysis,[15] a technique
which cleaves peptides bonds but does not cleave glycosidic
linkages except those to serine or threonine,[16] followed by
deacetylation, fractionation on Sephadex G-25 or G-150 and
sugar analysis indicates that the protein moiety was degraded to a molecular weight less than 500 and the carbohydrate side chains of glycoproteins from a variety of sources

Table 2. Amino acid composition of hydroxyproline-poor
glycoprotein from representatives of Pteridophyta,
Gymnospermae and Angiospermae.

Amino acid (Mol %)	Source						
	M. struthiopteris	P. stobus	N. odorata	V. faba	N. tabacum	R. rhaponticum	Z. mays
Hyp	trace	trace	trace	6	2	trace	trace
Asx	8	8	12	6	9	9	11
Thr	5	4	5	5	5	4	6
Ser	7	5	5	7	10	5	6
Glx	5	7	10	9	14	8	10
Pro	6	5	6	5	2	6	5
Gly	9	16	10	8	11	22	12
Ala	9	7	10	10	8	9	10
Val	8	5	10	7	7	7	7
Met	-	-	1	1	-	1	1
Ilc	6	4	5	4	5	4	5
Leu	10	6	8	10	7	7	8
Tyr	2	-	3	2	6	3	3
Phe	4	-	3	3	6	3	4
Tryp	-	4	2	3	trace	-	-
Lys	11	17	6	8	5	7	6
His	2	8	-	2	1	1	2
Arg	6	4	4	4	2	4	4

Table 3. Linkage analysis of hydroxyproline-poor glycoproteins from representatives of Pteridophyta,
Gymnospermae and Angiospermae.

Linkage (%)	Source						
	M. struthiopteris	P. stobus	N. odorata	V. faba	N. tabacum	R. rhaponticum	Z. mays
Ara (terminal	-	8	1	2	2	6	-
Ara (1 → 2)	-	-	-	-	-	11	-
Ara (1 → 3)	-	2	trace	-	trace	-	trace
Ara (1 → 5)	2	-	1	-	-	6	-
Xyl (terminal)	-	1	4	1	1	6	2
Xyl (1 → 4)	15	8	17	2	2	33	6
Gal (terminal)	6	4	-	-	-	13	-
Glc (terminal)	3	7	13	11	6	8	9
Glc (1 → 4)	75	70	63	84	89	17	81

were similar in molecular weight (Table 4). The optical
rotation of fractions containing predominately glucose indicate that this sugar is both α- and β-linked. Side chains
with a molecular weight greater than 2,000 were composed

Table 4. Molecular weight and sugar composition of oligo-saccharides obtained by trifluoroacetolysis of glycoproteins.

Fraction	Molecular Weight	Source	Percent (Weight)	Rha	Ara	Man	Gal	Glc	$[\alpha]\,^{D}_{20}$
1	> 2,000	a	33	10	28	8	27	28	-
		b	49	-	-	5	12	79	-
		c	34	-	2	-	-	98	-
		d	42	-	-	-	-	100	+ 111°
		e	34	-	-	-	-	100	+ 96°
		f	12	1	14	3	24	58	+ 159°
2	900-2,000	a	17	4	55	12	15	14	-
		b	18	-	30	9	16	45	-
		c	16	-	17	15	67	-	-
		d	40	-	-	5	13	82	+ 165°
		e	66	-	-	-	-	100	+ 96°
		f	83	3	10	4	16	67	+ 108°
3	500-900	a	14	-	48	18	18	15	-
		b	6	-	45	16	9	33	-
		c	15	-	25	28	10	38	-
		d	18	-	-	5	20	75	+ 180°
		e	-	-	-	-	-	-	-
		f	5	20	26	13	6	35	-
4	342-500	a	11	-	61	13	19	8	-
		b	11	-	43	10	15	29	-
		c	20	-	5	25	10	52	-
		d	-	-	-	-	-	-	-
		e	-	-	-	-	-	-	-
		f	-	-	-	-	-	-	-
5	180-342	a	24	-	47	14	34	5	-
		b	16	-	22	8	28	41	-
		c	16	-	44	-	56	-	-
		d	-	-	-	-	-	-	-
		e	-	-	-	-	-	-	-
		f	-	-	-	-	-	-	-

a. P.vulgaris hydroxyproline-rich glycoprotein.
b. P.vulgaris hydroxyproline-poor glycoprotein.
c. M.struthiopteris hydroxyproline-poor glycoprotein.
d. N.tabacum hydroxyproline-poor glycoprotein.
e. Z.mays hydroxyproline-poor glycoprotein.
f. R.rhaponticum hydroxyproline-poor glycoprotein.

mainly of glucose. Gel chromatography on Sephadex G-150 indicated the molecular weight of this fraction was heterogeneous with the largest fragments having molecular weights as high as 100,000. Side chains with a molecular weight

from 342 to 900 contained arabinose, galactose, glucose and
mannose, whereas, the fraction containing single sugar residues contained arabinose, galactose and, in the case of P.
vulgaris, glucose. No uronic acids were detected in any of
the fractions. Carbohydrate with a molecular weight greater
than 500 was pooled and placed on Sephadex DEAE (A50, acetate
form). Most of the carbohydrate was eluted with distilled
water. For example, 80% of the carbohydrate from N. tabacum
was eluted before the gradient of sodium acetate ($0 \rightarrow 2M$)
appeared in the effluent, providing further evidence that
the carbohydrate moiety of hydroxyproline-poor glycoprotein
does not contain uronic acids.

Hydroxyproline-poor glycoproteins from M. struthiopteris
and P. vulgaris both have N-terminal lysine indicating
further structural homology and suggesting that a hydroxyproline-poor glycoprotein with common structural features
is widely distributed among land plants. The isolation of
a proteoglycan from a red alga having an amino acid composition very similar to hydroxyproline-poor glycoprotein suggests that a glycoprotein with some structural similarities
also occurs among algae.[17]

Trifluoroacetolysis and gel chromatography of hydroxyproline-rich glycoprotein indicates that a significant portion of the carbohydrate moiety of this glycoprotein has a
molecular weight greater than 2,000 (Table 4). The high
optical rotation of glucose-rich fraction indicated most
glucose was α-linked ($[\alpha]_{20}^{D}$ + 180°; glucose, 65%; galactose,
13%; arabinose, 7% and mannose, 6%). Consequently, hydroxyproline-rich and hydroxyproline-poor glycoproteins both
appear to have ($1 \rightarrow 4$)-linked glucose residues most of which
are α-linked. Investigations to ascertain if this is a
structural feature of these glycoproteins or merely indicates a tightly-bound glucose polymer are currently being
initiated.

EXTRACTION OF ^{14}C-LABELLED GLYCOPROTEINS

The presence of hydroxyproline-poor glycoprotein in
all plants examined suggests an important function for this
glycoprotein in plant cells.[13] Knowledge of the cellular
location of this glycoprotein would aid elucidation of its
function. Addition of [^{14}C]proline to the Celite abrasive
used to wound leaf tissue of P. vulgaris,[5] labelled both

hydroxyproline-poor and hydroxyproline-rich glycoprotein
within 36 hours. Leaves were harvested 36 hours and 96 hours
after wounding and extracted by using one of three methods
which did not differ significantly from each other in the
extractants employed, but did differ in the order in which
the extractions were made. In the first method (Fig. 4),
cell walls were not treated with alkali before sodium
chlorite extraction. In the second method, hydroxyproline-
poor glycoprotein was extracted with cold alkali (2°C) before
preparation of cell walls and extraction with sodium chlorite.
In method three, cell walls were extracted with alkali (1M
and 4M at 20°C) after cell wall preparation but before
sodium chlorite extraction.

Radioactivity in each fraction was determined, after
drying, using a Packard Tricarb oxidizer. Supernatant fluids
were dialysed before freeze drying. Radioactivity in hydroxyproline and proline was determined after protein hydrolysis
using a 10:1 flow splitter during amino acid analysis so
that one-tenth of the column ouput went to the detector
while the remainder was collected for radioactivity measure-
ment. Radioactivity in proline and hydroxyproline accounted
for at least 70% of the radioactivity present in the samples
analysed. The ratio of hydroxyproline to proline is based
on the amount of radioactivity in these two amino acids and
is expressed as HP/P ratio.

Extraction by method 1 indicated that after 36 hours
proline was incorporated mainly into hydroxyproline-poor
protein that was extracted with sodium chlorite but was not
precipitated by ethanol (Table 5). Some proline was incorporated into hydroxyproline-rich glycoprotein particularly
if the specific activity of proline was low. Extending the
post wounding time from 36 hours to 96 hours before har-
vesting resulted in incorporation of proline into protein
which was associated with other cell wall fractions, mostly
the hemicellulose I fractions and cellulose. Extraction of
hydroxyproline-poor glycoprotein with cold alkali demonstrated that proline was incorporated into this glycoprotein;
however, most proline was incorporated into material which
occurred in the sodium chlorite supernatant fraction, but
this was also hydroxyproline-poor (Table 6). Extraction by
method 2 resulted in more incorporated proline being found
in other cell wall fractions, particularly the hemicellulose
I fractions, than method 1 (comparing results at 36 h).

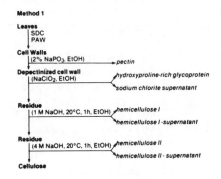

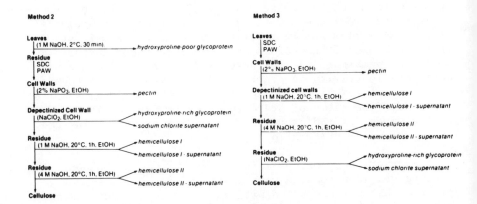

Figure 4. Extraction of leaves of P. vulgaris exposed to
[^{14}C]proline at wounding. Cell walls were prepared by a
modification of the method of Ring and Selvendran,[18] whereby
leaves were macerated in a Waring blender (10 min) followed
by a Polytron treatment (5 min) while suspended in sodium
deoxycholate (SDC, 1%). The insoluble material was collected on a sintered glass crucible and washed with SDC.
Maceration and washing were repeated twice. The residue
was suspended in phenol/acetic acid/water (PAW, 2:1:1),
macerated for 5 minutes in a Polytron, then collected by
centrifugation (5,000 g, 10 min). The PAW extraction was
repeated twice to yield cell walls.

Table 5. Incorporation of $[^{14}C]$proline into cell wall fractions of P. vulgaris (Method 1).

Fractions	Post Wounding Time							
	36 h				96 h			
	Expt. 1[a]		Expt. 2[a]		Expt. 1[a]		Expt. 2[a]	
	DPM	%	DPM	%	DPM	%	DPM	%
Hydroxyproline-rich glycoprotein[b]	10,267	7	603	-	2,252	18	4,035	12
Sodium chlorite supernatant[c]	144,108	93	1,015,800	100	3,349	27	6,412	20
Hemicellulose I	627	-	90	-	1,769	14	5,046	16
Hemicellulose I - supernatant	647	-	ND	-	761	6	5,000	15
Hemicellulose II	253	-	4,025	-	1,706	14	150	-
Hemicellulose II - supernatant	151	-	2,899	-	1,367	11	78	-
Cellulose	ND	-	3,936	-	1,432	11	11,785	36
Total	156,053		1,027,353		12,636		32,506	
Radioactivity used	0.8 μCi		8 μCi		0.8 μCi		0.8 μCi	

[a] Expt. 1 Specific activity of Proline 10 μCi/mMol
  Expt. 2 Specific activity of Proline 283 mCi/mMol
[b] HP/P = 1.2; HP/P is the ratio Hyp/Pro
[c] HP/P = 0.2

Table 6. Incorporation of $[^{14}C]$proline into cell walls fractions of P. vulgaris (Method 2).

Fractions	Post Wounding Time						
	36 h					96 h	
	Expt. 1		Expt. 2				
	DPM	%	DPM	%	HP/P	DPM	%
Hydroxyproline-poor glycoprotein	5,252	17	1,998	3	0.3	7,155	14
Hydroxyproline-rich glycoprotein	5,692	19	8,576	15	1.2	5,765	11
Sodium chlorite supernatant	12,725	42	28,000	48	0.3	13,590	26
Hemicellulose I	1,261	4	5,092	9	0.1	4,461	9
Hemicellulose I - supernatant	3,650	12	12,100	21	0.3	14,398	28
Hemicellulose II	361	1	1,320	2	-	213	-
Hemicellulose II - supernatant	100	-	0	-	-	5,551	11
Cellulose	982	3	1,210	2	-	1,038	2
Total	30,023		58,296			52,171	
Recovery (%)	90		79				

Specific activity of Proline: 10 μCi/mMol
Radioactivity used: 0.8 μCi
HP/P is the ratio Hyp/Pro.

Treatment with alkali at 20°C before sodium chlorite extraction altered the distribution of incorporated proline in the cell wall fractions (Table 7). Hydroxyproline-rich glycoprotein was no longer precipitated by ethanol and consequently it was present in the sodium chlorite supernatant fluid. This suggests that carbohydrate attached to serine may be required for ethanol to precipitate sodium chlorite extracted hydroxyproline-rich glycoprotein. Hydroxyproline-poor protein was associated with cellulose. These two fractions accounted for most of the proline incorporated into the cell wall. The recoveries of radioactivity given in Tables 6 and 7 are based on the amount of radioactivity present following hot water extraction in the hydroxyproline-poor glycoprotein extraction procedure (Fig. 2) for Table 6 and the radioactivity present in cell walls for Table 7. In each case, a sample was freeze-dried and combusted to CO_2 for radioactivity determination. A comparison of HP/P ratios at 36 hours and 96 hours post-wounding, indicates an increase in hydroxyproline content of all fractions with time.

A comparison of the distribution of incorporated [^{14}C]-proline in the fractions obtained by the three extraction methods (Table 8), indicated the following. Firstly, treatment with alkali increased the proportion of incorporated

Table 7.   Incorporation of [^{14}C]proline into cell wall
           fractions P. vulgaris (Method 3).

Fractions	Post Wounding Time 36 h			Post Wounding Time 96 h		
	DPM	%	HP/P	DPM	%	HP/P
Hemicellulose I	1,180	2		2,880	4	
Hemicellulose I supernatant	4,807	8	0.2	4,576	6	
Hemicellulose II	5,226	9	1.2	360	-	
Hemicellulose II - supernatant	8,436	14		693	1	0.8
Hydroxyproline-rich glycoprotein	544	1	0.9	2,049	3	1.5
Sodium chlorite supernatant	20,775	36	2.5	30,905	41	3.4
Cellulose	17,231	30	0.4	33,899	45	0.6
Total	58,199			75,362		
Recovery (%)	92			64		

Specific activity of Proline: 283 mCi/mMol
Radioactivity used: 0.8 μCi

proline found in hydroxyproline-rich fractions. This was particularly apparent when cell walls were treated with alkali at 20°C before sodium chlorite extraction (method 3 vs. method 2). The explanation of this observation may be related to differences in the extraction methods or to the observation that HP/P ratios were higher in all material obtained by method 3 than methods 2 or 1. This may indicate that, although all leaves were harvested at either 36 or 96 hours, biosynthesis of glycoproteins in leaves used for extraction by method 3 was more complete, consequently, more hydroxylation of proline residues had occurred. Secondly, treatment with alkali rendered hydroxyproline-poor protein more difficult to extract with sodium chlorite. Hydroxy-proline-poor glycoprotein was associated with hemicellulose after cold alkali treatment and with cellulose following treatment with alkali at 20°C. Again, this may be related to the extraction methods or differences in hydroxylation.

The designation of fractions as either hydroxyproline-poor or hydroxyproline-rich is arbitrary, that is, those fractions with a hydroxyproline/proline ratio less than 0.6 are termed hydroxyproline-poor; whereas, those fractions with a hydrosyproline/proline ratio greater than 0.6 are

Table 8. Distribution of [^{14}C]proline in hydroxyproline-poor and hydroxyproline-rich glycoproteins extracted by three methods.

	Percent Post Wounding Time					
	36 h			96 h		
Fractions	Method 1	Method 2	Method 3	Method 1	Method 2	Method 3
Hemicellulose I (Hyp-poor)	-	4	2	16	9	4
Hemicellulose I - supernatant (Hyp-poor)	-	12	8	15	28	6
Hemicellulose II - (Hyp-poor)	-	1	9	-	-	-
Hemicellulose II - supernatant	-	-	14	-	11	1
Hydroxyproline-rich glycoprotein	-	19	1	12	11	3
Sodium chlorite supernatant	100	42	36	20	26	41
	(Hyp-poor)	(Hyp-poor)	(Hyp-rich)	(Hyp-poor)	(Hyp-poor)	Hyp-rich)
Hydroxyproline-poor glycoprotein	-	17	-	-	14	-
Cellulose (Hyp-poor)	-	3	30	36	2	45

Method 1 - no alkali before sodium chlorite.
Method 2 - alkali (2°C) before sodium chlorite.
Method 3 - alkali (20°C) before sodium chlorite.

HP/P < 0.6 designated (Hyp-poor).
HP/P > 0.6 designated (Hyp-rich).

termed hydroxyproline-rich. The choice of 0.6 as the determining ratio is based on values obtained for a variety of preparations of hydroxyproline-poor and hydroxyproline-rich glycoprotein (Table 9).

Amino acid, sugar and linkage analysis verified the fraction termed hydroxyproline-rich glycoprotein was indeed this material. The results of linkage analysis was compared to a similar analysis of material from P. coccineus (Table 10). The glycoproteins were either prepared by chromatography on Sephadex G-200 (P. vulgaris) or DEAE-Sephadex (P. coccineus). The similarity of the results obtained is apparent despite the difference in preparative methods. The results are compatible with a structure containing tetra-arabinoside and terminal galactose which are characteristic of hydroxyproline-rich glycoprotein. One unexpected finding is the presence of (1→4)-linked glucose residues in both preparations.

SPECULATION

This chapter concludes with some speculation concerning the following questions. Firstly, what is the relationship between the hydroxyproline-poor glycoprotein extracted from wounded leaf tissue[4-8] and the hydroxyproline-poor protein found in cell walls?[1,11,12] Secondly, what relationship, if any, exists between hydroxyproline-poor and hydroxyproline-rich glycoprotein?

Table 9. Ratio of hydroxyproline to proline in preparations of hydroxyproline-rich and hydroxyproline-poor glycoproteins from Phaseolus.

Glycoprotein	HP/P	Source	Reference
Hydroxyproline-rich glycoprotein	0.7	P. vulgaris	Lamport, 1965
	1.1	P. vulgaris	Brown and Kimmins, in press
	3.0	P. vulgaris	Brown and Kimmins, in press
	7.0	P. coccineus	Selvendran, 1975
	7.3	P. coccineus	O'Neill and Selvendran, 1980
Hydroxyproline-poor glycoprotein	0.3	P. vulgaris	Brown and Kimmins, 1978
	0.2	P. vulgaris	Brown and Kimmins, 1978
	1.0	P. vulgaris	Brown and Kimmins, 1978
	0.3	P. vulgaris	Brown and Kimmins, in press
	0.5	P. coccineus	Selvendran, 1975

Table 10. Linkage analysis of hydroxyproline-rich glycoprotein from P. vulgaris and P. coccineus.

	Mol (%)	
Linkage	P. vulgaris[a]	P. coccineus[b]
Rham (1 → 2)	-	1
Ara (terminal)	16	23
Ara (1 → 2)	24	35
Ara (1 → 3)	12	19
Ara (1 → 5)	2	3
Xyl (terminal)	3	1
Xyl (1 → 4)	-	2
Gal (terminal)	5	9
Gal (1 → 4)	7	1
Glc (terminal)	2	-
Glc (1 → 4)	27	5

[a]Brown and Kimmins (in press): Extraction; Selvendran, 1975:
purified by chromatography on Sephadex G-200.

[b]O'Neill and Selvendran, 1980: Extraction; modified
Selvendran, 1975: purified by chromatography on DEAE-Sephadex.

Investigation of the first question depends on the
presence of a characteristic structural feature which can
be demonstrated to be common in hydroxyproline-poor glycoprotein from wounded leaf tissue and cell walls. Unlike
hydroxyproline-rich glycoprotein which has a readily detectable alkali-stable hydroxyproline-O-arabinosyl linkage, no
characteristic structural feature of hydroxyproline-poor
glycoprotein is known at the present time. However, alkali
treatment (0.2 M NaOH, 50°C, 5 h) produces at least three
peptide fragments having proline and glycine as N-terminal
units, in addition to lysine, which is the N-terminus of
hydroxyproline-poor glycoprotein.[8,13] These polypeptide
fragments have molecular weights of 143,000, 138,000, and
11,000, determined by PAGE electrophoresis and column
chromatography. Addition of sodium [^{35}S]sulphite during
alkali treatment labels only the smallest peptide fragments
(11,000) indicating that most carbohydrate attached to serine
is associated with these fragments.[8] Investigations to determine if these structural features were present in cell
wall hydroxyproline-poor glycoprotein were complicated by
the finding that sodium chlorite-extracted hydroxyproline-

rich glycoprotein yielded ^{35}S-containing material having a molecular weight of 17,000 following alkali treatment.

Alkali treatment in the presence of sodium borohydride (1 M) produced fragments from M strupthiopteris and R. rhaponticum hydroxyproline-poor glycoproteins which isoelectric focusing (48 h, pH 3-10) indicated had the following isoelectric points; 3.85, 4.15, and 4.3 (M. strupthiopteris), 3.0 and 3.4 (R. rhaponticum). Gel electrophoresis indicated molecular weights in the range 650 to 1300. Material not soluble in the light solution of amphophytes was removed by centrifugation before isoelectric focusing. Present investigations seek to compare peptide fragments produced from leaf and cell-wall hydroxyproline-poor glycoproteins following treatment with alkali to determine if any homology exists between the two proteins.

Linkage analysis of hydroxyproline-rich glycoprotein after alkali treatment indicated that this procedure resulted in loss of (1→4)-α-linked glucose units and an increase in terminal, (1→3)- and (1→5)-linked arabinose units as well as terminal galactose (Table 11). The latter result was unexpected as hydroxyproline-rich glycoprotein has single galactose residues attached to serine,[19] however, similar treatment of potato lectin released galactose residues from serine very slowly.[20] This was attributed to an inhibitory effect of the arabinofuranosidic residues on the β-elimination reaction caused by a negative charge on the hydroxy groups of adjacent arabinofuranosidic residues. If this explanation is correct, some (1→2)-linked arabinose units and (1→4)-linked glucose units attached to serine must be sufficiently removed from hydroxyproline-tetraarabinoside units so that this inhibition does not occur.

The occurrence of hydroxyproline-poor and hydroxyproline-rich glycoprotein in cell walls poses the question of what relationship, if any, exists between hydroxyproline-poor and hydroxyproline-rich glycoprotein? The possibilities include the following; (1) hydroxyproline-rich glycoprotein is a part of hydroxyproline-poor glycoprotein which is separated by the extractants employed in its isolation or as a normal development process during cell maturation, (2) hydroxyproline-poor glycoprotein is a precursor of hydroxyproline-rich glycoprotein and (3) the two glycoproteins are synthesized independently. Although there is little, if

Table 11. Linkage analysis of hydroxyproline-rich glycoprotein before and after β-elimination.

Linkage	Mol (%)	
	Before β-elimination	After β-elimination
Ara (terminal)	16	23
Ara (1 → 2)	24	27
Ara (1 → 3)	12	23
Ara (1 → 5)	2	11
Xyl (terminal)	3	5
Gal (terminal)	5	11
Gal (1 → 4)	7	-
Glc (terminal)	2	-
Glc (1 → 4)	27	-

any, information which supports any of these possibilities,
an attractive hypothesis results from a combination of the
first two, whereby, some proline residues of hydroxyproline-
poor glycoprotein would be oxidized and glycosylated to
yield regions high in hydroxyproline. Cleavage of these
regions (enzymic or chemical) could release a hydroxyproline-rich glycoprotein. The two glycoproteins have common
structural features including (1→2)-, (1→3)-, and (1→5)-
linked arabinose units, (1→4)-linked glucose residues, and
terminal galactose.

Hydroxyproline-poor glycoprotein binds to cellulose,
agarose and in one case, dextran. These polysaccharides
have no obvious structural features in common, therefore,
binding to the polysaccharides may involve different regions
of hydroxyproline-poor glycoprotein. Lamport has proposed
that hydroxyproline-rich glycoprotein has a hydrophobic tail
and hydrophilic head.[21] Thus, the larger complex, (hydroxyproline-poor glycoprotein) may have many regions which bind
to a variety of material. Treatment of agarose-bound
hydroxyproline-poor glycoprotein with alkali-sodium sulphite
releases a fragment high in cysteic acid,[7] suggesting that
carbohydrate attached to serine is responsible for this
binding. Labelled hydroxyproline-poor glycoprotein is associated with hemicellulose if treated with cold alkali and
with cellulose if treated with alkali at room temperature
(Table 8) suggesting that this glycoprotein is associated
with several cell wall components. The hypothesis proposed

is that hydroxyproline-poor glycoprotein associates with
many cell wall components and selective cleavage of this
glycoprotein could release cell wall moieties from the
matrix in a prescribed manner. Several studies have demonstrated that an increase of cell wall hydroxyproline relative
to cell wall proline is correlated with cessation of cell
elongation.[22,23] In the model we propose (Fig. 5), hydroxyproline-poor glycoprotein would bind to a cell wall component,
probably cellulose; then, hydroxylation of proline residues
would occur as the cell matures. Ultimately, a hydroxyproline-rich glycoprotein moiety would be generated which may
remain with the parent molecule or be cleaved enzymically
to release a cell wall component whose role at the present
time is not understood.

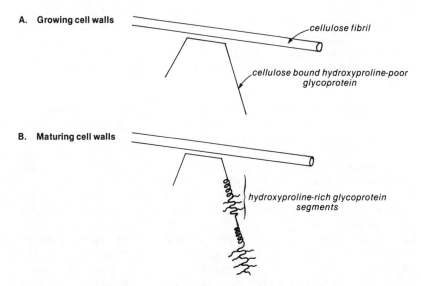

Figure 5. Postulated structure of hydroxyproline-poor glyco-
          protein complex in growing and non-growing cell
          walls.

ACKNOWLEDGEMENTS

The authors thank Mr. Keith Long, Mrs. A. Schramm and Miss D. Devlin for technical assistance, Dr. M. H. V. Laycock for N-terminal amino acid analysis, and the Natural Sciences and Engineering Research Council Canada for financial support.

REFERENCES

1. Selvendran, R. R. 1975. Cell wall glycoproteins and polysaccharides of parenchyma of Phaseolus coccineus. Phytochemistry 14:2175-2180.
2. Lamport, D. T. A. 1965. The protein component of primary cell walls. In (R. D. Preston, ed.), Adv. Bot. Res. 2:151-218.
3. Roberts, K. 1979. Hydroxyproline: its asymmetric distribution in cell wall glycoprotein. Planta 146:275-279.
4. Brown, R. G. and W. C. Kimmins. 1973. Hypersensitive resistance. Isolation and characterization of glyco-proteins from plants with localized infections. Can. J. Bot. 51:1917-1922.
5. Kimmins, W. C. and R. G. Brown. 1973. Hypersensitive resistance. The role of cell wall glycoproteins in virus localization. Can. J. Bot. 51:1923-1926.
6. Brown, R. G., W. C. Kimmins and B. Lindberg. 1975. Structural studies of glycoproteins from Phaselous vulgaris. Acta. Chem. Scand. B29:843-852.
7. Brown, R. G. and W. C. Kimmins. 1978. Protein-poly-saccharide linkages in glycoproteins from Phaselous vulgaris. Phytochemistry 17:29-33.
8. Brown, R. G. and W. C. Kimmins. 1979. Linkage analy-sis of hydroxyproline-poor glycoprotein from Phaselous vulgaris. Plant Physiol. 63:557-561.
9. Monro, J. A., R. W. Bailey and D. Penny. 1974. Cell wall hydroxyproline-polysaccharide associations in Lupinus hypocotyls. Phytochemistry 13:375-380.
10. Monro, J. A., R. W. Bailey and D. Penny. 1975. Differential alkali-extraction of hemicellulose and hydroxyproline from non-delignified cell walls of lupin hypocotyls. Carbohyr. Res. 41:153-161.
11. O'Neill, M. A. and R. R. Selvendran. 1980. Glycoproteins from the cell wall of Phaseolus coccineus. Biochem. J. 187:53-63.

12. Kimmins, W. C., R. G. Brown and K. Nowski. Incorporation and distribution of ^{14}C-proline in cell walls of bean. Can. J. Bot. (In press).
13. Brown, R. G. and W. C. Kimmins. 1980. Hydroxyproline-poor glycoprotein: its occurrence among land plants. Can. J. Bot. (In press).
14. Lamport, D. T. A. and D. H. Miller. 1971. Hydroxyproline arabinosides in the plant kingdom. Plant Physiol. 48:454-456.
15. Nilsson, B. and S. Svensson. 1979. Studies of the reactivity of methylglycosides, oligosaccharides and polysaccharides toward trifluoroacetolysis. Carbohydr. Res. 69:292-296.
16. Lindberg, B., B. Nilsson, T. Norberg and S. Svensson. 1979. Specific cleavage of O-glycosidic bonds to L-serine and L-threonine by trifluoroacetolysis. Acta Chem. Scand. 33B:230-231.
17. Court, G. J. and I. E. P. Taylor. 1979. Isolation and partial characterization of a proteoglycan from the red alga Laurencia spectabilis. Phytochemistry 18:411-414.
18. Ring, S. and R. R. Selvendran. 1978. Purification and methylation analysis of cell wall material from Solanum tuberosum. Phytochemistry 17:745-752.
19. Lamport, D. T. A., L. Katona and S. Roerig. 1973. Galactosylerine in extensin. Biochem. J. 133:125-131.
20. Allen, A. K., N. N. Desai and A. Newberger. 1978. Properties of potato lectin and the nature of its glycoprotein linkages. Biochem. J. 171:665-674.
21. Lamport, D. T. A. 1977. Structure, biosynthesis, and significance of cell wall glycoproteins. Recent Adv. Phytochemistry 11:79-115.
22. Bailey, R. W. and H. Kauss. 1974. Extraction of hydroxyproline containing proteins and pectic substances from cell walls of growing and non-growing mung bean hypocotyl segments. Planta 119:233-245.
23. Sadava, D., F. Walker and M. J. Chrispeels. 1973. Hydroxyproline-rich cell wall protein (extensin): Biosynthesis and accumulation in growing pea epicotyls. Dev. Biol. 30:42-48.

Chapter Seven

PLANT PROTOPLAST AGGLUTINATION AND IMMOBILIZATION

PHILIP J. LARKIN

Division of Plant Industry, CSIRO
Canberra City, A.C.T. 2601, Australia

INTRODUCTION

It is now an almost ubiquitous notion amongst plant cell biologists that plant cells do possess recognition faculties. We have in mind models which envisage plant cells regulating some of their functions in response to external stimuli. These external stimuli may also be thought of in terms of communicated information such as in pollen/stigma inter-actions or pathogen elicitor induction of host phytoalexins.

The models for recognition processes generally include lock and key (antibody/antigen or enzyme/substrate) type interactions between the recognition binder and the informa-tion determinant. Many of the models also assume that the information for communication is encoded by carbohydrate moieties. This is so partly because lectins are currently so newsworthy.[1-3] Additionally, however, one may argue that carbohydrate moieties have a far greater information carrying capacity than a peptide with the same number of residues. This results from the fact that the polysaccharide may be branched, the glycosides linked through different positions and by either α- or β-linkages. Also the resulting moiety is spatially rigid allowing a 3-dimensional "key" to be constructed to fit a given "lock". This trend to model

information determinants as carbohydrates has a conceptual
drawback: polysaccharides are at least two steps removed
from the genetic code. The central dogma requires us to see
all heritable information as coded in nucleic acid. We do
not yet fully understand how 3-dimensional polysaccharide
information is faithfully translated from the genetic code.
Nevertheless cells can achieve this as illustrated by the
simple inheritance of the ABO blood group determinants.

Now a conceptual stumbling-block for such models is the
location of the binders (recognition molecules) and determinants. The plant cell wall is an attractive location because complementary binders or determinants from the external
milieu have unimpaired access to the wall. However it is
difficult to conceive how a recognition interaction occurring
on the external extremities of the wall can have an effect
on the more active machinery of the cell. A plasmalemma
location for the critical interaction is much more attractive
in terms of switching on or off cell functions. However one
can not help but be a little concerned that the wall may be
an obstacle for some determinants arriving at the critical
site. This conceptual difficulty would seem most acute
where the recognition phenomenon is between two plant cells
where the determinant and binder are both anchored in the
plasmalemma.

These difficulties of location are sometimes solved for
us by the details of a recognition system. For example in
many fungal pathogenic systems the hyphae penetrate the wall
and sometimes can even be seen to bind to the host plasmalemma.[4,5] Often a fungal haustorium develops between the
host wall and plasmalemma.[6,7] In powdery mildew of barley
the compatible or incompatible relationship seems to be
determined when penetrating hyphae contact the host plasmalemma.[8] Tomiyama et al.[9] have observed that Phytophthora
infestans hyphae initiate potato hypersensitive reactions
only upon contact with the host plasmalemma. These authors
also showed that fungal extracts increase potato protoplast
death only in compatible combinations.

Another fascinating example where the location problem
seems to be solved for us is in graft compatibility phenomena.
Where the dividing cells of scion and stock come together the
opposing cell walls are dissolved and the plasmalemmas of the
two cell types come into direct contact.[10]

These examples not only implicate the plasmalemma as the primary location of the critical recognition interaction but also illustrate some ways in which the "obstruction" of the wall is overcome. It must also be said that our impulsive conception of the membrane and wall being discontinuous and divorced from each other is probably quite wrong. Roland[11] has reviewed the evidence for the close relationship and interdependence of the plant cell wall and membrane. Even ultrastructurally it is apparent that fibrils anchored in the membrane extend deep into the wall. There are also pro-toplasmic outgrowths and plasmalemma extensions that can be visualized in the wall. It is quite likely that some poly-saccharides or glycoproteins anchored in the membrane may extend right through the wall to expose determinants beyond the wall.

The likelihood that membranes are the primary sites of recognition events is enhanced by the observation that many plant lectins appear to be localized to membranes.[12-17] However some have also been localized in the cytoplasm,[14,19] and some to cell walls.[17,18]

Isolated plant protoplasts are cells lacking a cell wall with the plasmalemma directly exposed to the external milieu. Protoplasts may offer advantages as an experimental approach to the study of recognition interactions which occur at the plasmalemma. The preparation of protoplasts is perhaps the only reliable means of obtaining a suspension of truly iso-lated single cells from a given plant tissue. As a conse-quence of these characteristics, it is proposed that proto-plast agglutination or immobilization may be used as indi-cators of certain recognition events. Our consideration of these possibilities will be categorized on the basis of whether the determinant and binder are fixed (e.g. anchored to the plasmalemma) or free (e.g. solubilized from the cell).

PROTOPLAST AGGLUTINATION

Determinant fixed/ binder free

The naked plasmalemma of a protoplast will have its determinants entirely exposed. If a multivalent binder for one or more of these determinants is added exogenously to the protoplast suspension the interaction can manifest itself by protoplast agglutination. This has been demonstrated for

a number of isolated plant lectins with many species of
protoplasts (Table 1, Fig. 1).

Table 1. Exogenous lectin-mediated agglutination

Lectin source	Lectin	Agglutin- ation	No. of protoplast species	Reference
Arachis hypogaea (peanut)	PNA	+	8	23, 25
Glycine max (soybean)	SBA VII	+	11	23, 25, 26
Canavalia ensiformis	ConA	+	13	20-24, 26, 27
Ricinus communis (castor bean)	RCA II	+	8	23, 26
Triticum aestivum (wheat germ)	WGA	+ −	4 7	24 23
Phaseolus vulgaris (red kidney bean)	PHA-M or-P	+ −	4 7	24 23, 26
Ulex europaeus	UEA	−	7	23, 26
Bandeiraea simplicifolia		−	5	23
Dolichos biflorus	DBA	−	4	23, 26
Phytolacca americana (pokeweed)	PWM	−	1	26

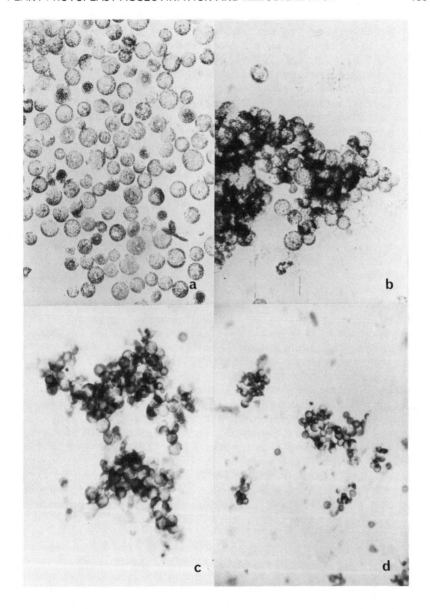

Figure 1. Agglutination induced by exogenous soluble lectins.
Petunia hybrida leaf protoplasts.   a) Control (x100).
b) Agglutination with PNA (x100).
    Triticum aestivum leaf protoplasts.   c) Agglutination
with SBA VII (x80).   d) Agglutination with RCA II (x50).

Multivalency (more than one binding site per molecule) is a critical factor in allowing the crosslinking between cells which leads to agglutination. Some of the well-characterized lectins require pH to be raised to about 7 before they form the multimers which are most efficient at cell agglutination.[2,28] The ability to agglutinate can be enhanced even further by cross-linking the lectin molecules with agents such as glutaraldehyde.[29] When attempting to use protoplast agglutination to investigate undefined receptors one needs to satisfy the requirement for multivalency either by varying parameters such as pH or crosslinking or covalently attaching the binders to a matrix (see later under protoplast immobilization).

The ability to agglutinate will also probably be dependent upon the density or mobility of the determinants on the protoplast surface. Effective bridging between cells may require many lectins (recognition molecules) to bind at any one contact site. This in turn may require the determinants to be mobile and "cap" at the site to give the required local density. Conceivably some determinants may be relatively immobilized due to functional relationships with certain structures. This inability to cap may prevent their use as receptors mediating protoplast agglutination.

The lectins reported to cause agglutination of plant protoplasts represent different glycoside specificities. The agglutination is inhibited by the appropriate simple sugars or glycosides.[20,23,24] This confirms that it is a function of the lectin activity. The lectins do not differentiate between plant species in the ability to bind to their protoplasts. This is perhaps no surprise. Legume lectin-mediated agglutination may be an entirely artificial phenomenon and the protoplast agglutination per se may be unrelated to any in vivo function of these seed lectins.

There is now a widespread consensus that legume lectins play a determining role in the specificity of Rhizobium-legume symbiosis,[30-41] although not all reports have concurred.[42,43] Most of this work has concerned the interaction of the plant lectin with the rhizobial cells. Dazzo and Hubbell[33] proposed a model for attachment based upon a cross-reactive antigen on both the Rhizobium and plant cell. The recognition lectin is proposed to bind to this antigen and, due to its multivalency, form a bridge attachment between the

rhizobial cells and plant root hairs. There has however
been very little work to confirm the existence of cross-
reactive antigens. If they are anchored in the host plasmalemma then protoplast preparations may be agglutinated by
the appropriate lectin.

Some toxins are known to have subunits which have lectin
activity. Cholera toxin,[44] tetanus toxin,[45] and diphtheria
toxin[46] are able to specifically bind to cell surface gang-
liosides or oligosaccharides. There may be toxins from plant
pathogens which similarly have lectin-type activities. If
these are multivalent, or can be made multivalent, with
respect to that activity (e.g. by glutaraldehyde cross-
linking) this surface interaction may be manifested as an
ability to agglutinate protoplasts. Protoplast agglutination would then become a system for investigating the host-
specificity and chemistry of the interaction.

Yeoman et al.[10] have reported agglutination of the
protoplasts of one species caused by saline leaf extracts of
another species when the two species are graft compatible.
No agglutination was observed in graft incompatible combinations. These preliminary observations need extension and
clarification. As they are they seem to indicate that the
protoplast agglutination system has uncovered the recognition components involved with graft compatibility.

Mention should also be made of the use of antisera to
agglutinate protoplasts. This is of course not a plant
recognition system. However antibody-mediated agglutination
is a potentially useful tool for confirming the plasmalemma-
location of a given component. A number of reports have confirmed that antiserum causes protoplasts to aggregate.[47-51]
However even control sera were found to cause high levels of
agglutination apparently due to the interaction of a non-
specific component of the serum of 9 animal species with protoplast surface arabinogalactan proteins[50] (Fig. 2). Partial
purification of the immunoglobulin by ammonium sulphate fractionation did not remove the non-specific agglutinins.

Recently Raff et al.[51] confirmed that control (pre-
immune) rabbit sera cause plant protoplast agglutination.
Indeed they were using the IgG fraction prepared by Protein
A-Sepharose affinity chromatography. At low IgG concentrations there appeared to be a greater ability to agglutinate

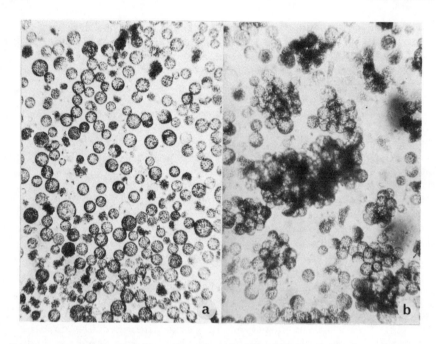

Figure 2. Agglutination with pre-immune immunoglobulins.
Nicotiana tabacum leaf protopolasts.   a) Supernatant of 45%
saturation $(NH_4)_2SO_4$ fractionation of cat serum.   b) Precipitate (immunoglobulins) of 45% saturation $(NH_4)_2SO_4$ fractionation of cat serum.

Both fractions were dialyzed to protoplast osmoticum and
used at a concentration of 1/10 relative to original volume.

with the immune preparations.   Nevertheless there was 75% agglutination when the control preparation was used at 5 µg/ml.
Hanke[25] observed agglutination triggered by submaxillary
asialomucins.   Similar components in serum may also bind to
Protein A.   There is a need to clarify which serum component
is involved in the non-specific agglutination.   If it proves
possible to prepare highly purified immunoglobulin specific
for surface antigens they will be useful for confirming the
observations of Strobel and Hess[49] that the recognition protein of sugarcane for the Helminthosporium sacchari toxin
exists in the protoplast plasmalemma.

Determinant free/binder fixed

The naked surface of a protoplast will have any plasma-
lemma-bound recognition binding molecules exposed for easy
access of exogenous determinants. If the determinants are
multiliganded the interaction may manifest itself as protoplast agglutination. This category of interaction may be
illustrated with the agglutination induced by Yariv antigens.
These are synthetic phenylazo-glycosides which are trivalent
with respect to the glycoside moiety[52,53] (Fig. 3). These
synthetic determinants, when bearing suitable β-linked
glycosides, are bound by arabinogalactan proteins (sometimes
referred to as β-lectins) on the protoplasts and agglutination results[50,54] (Fig. 4). It is not known what function
the arabinogalactan proteins serve in plants. Gleeson and
Clarke[55] found them as major components of stylar canals and
suggested a recognition and/or nutritive role for the growing
pollen tube. It has also been suggested that they may be
pollen catching molecules in the stigma.[56] The structure
and function of arabinogalactan proteins has recently been
reviewed by Clarke et al.[57]

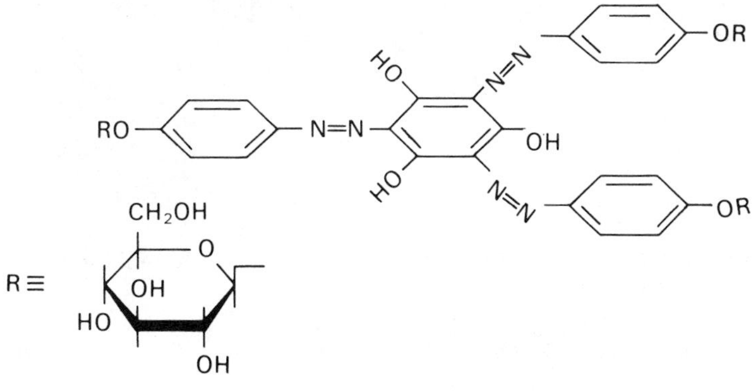

Figure 3. Structure of Yariv antigens. Shown is the β-
glucosyl Yariv (β-D-Glu) which is 1,3,5-tri-(p-β-D-glucosy-
loxyphenylazo)-2,4,6-trihydroxybenzene.

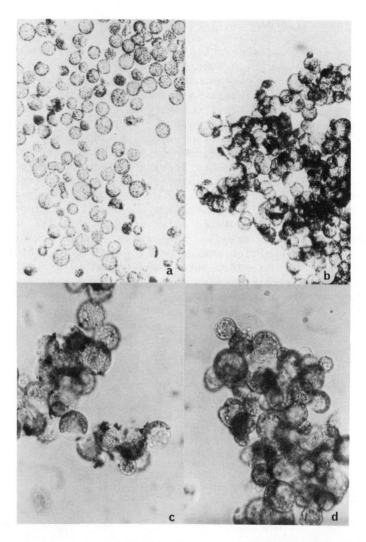

Figure 4. Agglutination of protoplasts induced by β-glycosyl
Yariv antigens.
    Petunia hybrida leaf protoplasts.   a) Control (x100).
b) Agglutination with 50 μg/ml β-lactosyl Yariv antigen
(x100).
    Triticum aestivum leaf protoplasts.   c) Agglutination
with 50 μg/ml β-galactosyl Yariv antigen (x250).   d) Agglutination with 50 μg/ml β-cellobiosyl Yariv antigen (x250).

It is also interesting that arabinogalactans themselves
are good determinants for other lectins, possibly due to
their brush-like branching structure.[58] They have been used
in cross-linked form as affinity matrices for the purification of castor bean and peanut lectins.[59,60] It seems quite
likely that surface arabinogalactans are at least some of
the determinants acting as receptors during RCA II, PNA- and
SBA VII-induced protoplast agglutination.

It is a possibility that phenolic glycosides such as
flavonoids are the natural analogues of Yariv antigens.
There is evidence that O-hydroxymethylphenyl-β-$\underline{D}$-glucoside
(salicin), p-nitrophenyl-β-$\underline{D}$-glucoside and indoxyl-β-$\underline{D}$-
glucoside are inhibitory to Yariv antigen-mediated agglutin-
ation.[54] Some partially purified flavonoid glycosides extracted from Ceratonia siliqua (carob), Hypericum perforatum,
Echeveria gibbiflora and Pseudotsuga menziersii (Douglas-fir)
leaves and Allium cepa (onion) bulbs also showed inhibitory
activity.[61] Jermyn[62] has found that β-lectins are retained
on affinity columns of myricetin-3-β-$\underline{D}$-galactoside or
quercetin-3-β-$\underline{D}$-galactoside coupled to the bromoacetyl
derivative of 6-amino-hexyl Sepharose. These flavanol
glycosides also competitively inhibited the precipitation
reaction between β-lectins and Yariv antigens.

A tantalising example of protoplast agglutination which
has bearing on the theme of recognition is that reported by
Professor Stelzig and his colleagues.[63,64] They find that
the elicitor from Phytophthora infestans which induces
phytoalexin production in potato discs also causes extensive
agglutination of potato leaf protoplasts. The elicitor is a
β-1,3-glucan with extensive 1,6-branching. The agglutination
is taken as corroborative evidence that there is a recognition receptor on the host plasmalemma.

There are of course many other elicitor/phytoalexin
systems. In most of these the elicitor is reported to be
non-host-specific.[65-70] Keen[71,72] however has obtained a
mannose-containing, wall-associated glycoprotein elicitor
from Phytophthora megasperma var. sojae which appears to
be host-specific.

Similarly there are race-specific glycoproteins from
Cladosporium fulvum which induce hypersensitive cell death
only in resistant tomato hosts.[73,74] It may prove valuable

to investigate the ability of these elicitors to agglutinate
protoplast preparations. One could vary the genotype of the
protoplasts to determine the host range for receptors
(binders) of the elicitors.

It is interesting that a terpenoid glycoside with
elicitor activity appears to have been recovered as a sub-
fraction of P. infestans wall elicitors.[66] In potato these
elicit phytoalexins which are themselves terpenoids. This
observation requires further clarification. It is possible
that arabinogalactan proteins (β-lectins) are the binding
receptors for terpenoid elicitor moieties.

More recently hypersensitivity-inhibiting-factors (HIF)
have been described in the potato/Phytophthora infestans
system.[75] These suppressors, like the elicitors themselves,
are glucans containing $\beta(1\rightarrow3)$ linkages. They are isolated
from the fungal mycelia, zoospores and cytospores as small
molecular weight water-soluble molecules. The glucans from
compatible races have greater suppressor activity than those
from incompatible races and it is suggested that these
factors are responsible for the specificity. It is likely
they act by competitive inhibition of the elicitor interaction with the plant cell receptor. If this mode of action
is correct they should also inhibit potato protoplast
agglutination in the presence of elicitor.

There are other examples of pathogenic toxins which
appear to induce susceptibility rather than resistance (as
in the case of hypersensitivity elicitors). El-Banoby and
Rudolph[76] recently described water-soaking extracellular
polysaccharides from Pseudomonas and Xanthomonas pathogens
which appear to be specific for their respective hosts.
The mal-secco disease of Citrus sp. is caused by Phoma
tracheiphila and the pathogen produces a glycoprotein toxin
(malseccin) which plays a role in the disease process.[77]
Such toxins may also have specific binding receptors on host
plasmalemma which may be indicated by protoplast agglutination in their presence. The host-specific toxin, victorin,
from Helminthosporium victoriae causes susceptible oat protoplast lysis.[78] Cross-linking of the toxin into multimers
may allow oat protoplast agglutination and possibly suppres-
sion of the lysis effect.

Determinant fixed/binder fixed

Many or most recognition phenomena will involve both
determinant and binder fixed to separate cells. When one
of these components can be solubilized the experimental
possibilities discussed above apply. We will consider here
also systems where neither determinant nor binder can be or
need be solubilized. The phase-, sex-, and species-specific
mating type reaction in Chlamydomonas is a naturally-occur-
ring agglutination system.[79] The initial step in the copu-
lation process is the adhesion of the flagella of opposite
mating types of gametes of the same species. The adhesion
appears to involve two independent binding phenomena: a
binder on the + gamete with a determinant on the - gamete,
and a binder on the - gamete with a determinant on the +
gamete. All components in the interaction are fixed to the
cell surfaces. It is possible to solubilize agglutinin
activities with the predicted specificities. However the
mating interaction can be observed as agglutination of +
and - gametes under experimental conditions with all components in situ.

It may be possible to employ this approach with mix-
tures of plant protoplasts and pathogenic cells. If there
is a binding-type recognition interaction between the two
cell types, then a coagglutination matrix will form. There
is considerable evidence that, as a prerequisite for plant
tumor induction, the bacterium Agrobacterium tumefaciens
attaches to specific sites on the plant cell wall.[80-84] All
cell wall materials are synthesized and extruded through the
plasmalemma. It is therefore possible that the protoplast
will also bear the receptors for agrobacterial adhesion. A
few hours preincubation in culture medium allowing synthetic
activity may enhance this possibility.

Similar experimental possibilities exist with the
hypersensitivity-inducing bacterium, Pseudomonas solanacearum.
The hypersensitive response of the plant tissue is triggered
by incompatible bacterial strains which are "rough" in lipopolysaccharide type. The initial event which triggers the
response seems to be attachment of the bacteria via the
lipopolysaccharide determinants to a plant cell surface
lectin.[85-88] Solubilized potato lectin appears to have the
required selectivity between virulent and avirulent strains.[83]
Bacterial immobilization by plant cell surface components

has been postulated as a defence mechanism in a number of
other systems as well.[89-92]

In many fungal pathogen interactions, the critical
recognition event appears to occur when the penetrating
hyphae contacts the plant plasmalemma.[6-8] Lectin-like
activities have been implicated.[67] Hyphal fragments may
bridge between appropriate plant protoplasts causing
agglutination.

Reference has already been made to the cell contact
between stock and scion in a plant graft. The cell walls
of the dividing cells appear to be digested at the point of
contact to allow direct contact between the plasmalemmas of
the stock- and scion-derived cells.[10] This plasmalemma
exposure and contact occurs both in compatible and incompatible grafts and it is only after this phenomenon that
incompatibility becomes apparent. Protoplasts may be able
to make a valuable contribution to the study of graft compatibility. Do protoplasts prepared from the stock callus
and scion callus autoagglutinate when mixed? Preliminary
observations with seed extracted agglutinins and leaf protoplasts showed selective agglutination with graft-compatible
partners.[10] This work needs extension and clarification.
There may be specific recognition binders and determinants
produced at the time of graft contact.

Protoplast immobilization

An alternative experimental possibility for using
protoplasts to investigate surface recognition events is to
artificially fix one of the components (binder or determinant) to a solid support. If a solubilized determinant is
uniliganded, or a solubilized binder is univalent, they will
be unable to initiate protoplast agglutination. If such
components are first covalently linked to a solid matrix,
the recognition interaction will be visualized as protoplast
immobilization. We have been able to demonstrate this approach using Con A, RCA II, SBA VII, and PNA covalently
bound with glutaraldehyde to collagen membranes or serum
albumin sponges (themselves formed with glutaraldehyde
cross-linking). Plant protoplasts adhere tenaciously to
such surfaces so that they are stable to thorough rinsing
and viable at least in the short term (Fig. 5).

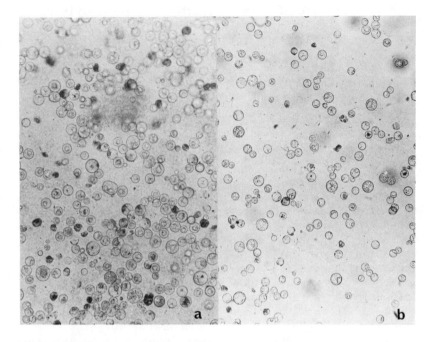

Figure 5. Protoplast immobilization. a) Tobacco "nia 63"
culture protoplasts adhering to a glutaraldehyde cross-
linked protein sponge containing bovine serum albumin and
Con A. The non-adhering protoplasts have been removed by 3
consective rinses by immersion. b) Corn culture protoplasts
immobilized similarly to above.

        This approach, while technically more involved, has
the advantage that components of unknown valency may be used.
It may also be possible to differentiate between biological
effects requiring uptake of the effector and effects requiring only surface binding. For example pathotoxins or
phytohormones may be bound to a supporting matrix and plant
protoplasts immobilized by interaction with the toxins or
phytohormones. If the biological effects can still be observed it may be concluded that uptake of the effector is
unnecessary.

A modification of this approach is to covalently cross-link effectors (determinants or binders) to erythrocytes or beads. The recognition interaction is then visualized as a rosetting of the erythrocytes or beads around the plant protoplasts. Raff et al.[51] have shown that the rosetting technique can be used to confirm the location of an antigen on plant protoplasts. The protoplasts were reacted with the rabbit antiserum. Erythrocytes bearing cross-linked sheep anti-rabbit immunoglobulin formed rosettes around the protoplasts.

CAUTION AND CONCLUSION

Anyone contemplating the use of protoplast agglutination to investigate some recognition phenomenon needs to be aware of the dangers. Firstly one must be aware that there are a number of macromolecules which cause non-specific agglutination of protoplasts. Those reported include gelatin proteins and peptides,[93] non-immune immunoglobulin preparations,[50, 51] IgA myeloma J539 protein,[94] submaxillary asialomucins,[25] polyethylene glycol,[95, 96] and polyvinyl-alcohol.[97] It is also noteworthy than some α-galactosidases extracted from legume seeds acted as haemagglutinins.[98] Some species of protoplasts we have observed to be particularly prone to autoagglutination particularly after a few hours in certain culture media e.g. Brassica protoplasts in B5 salt[99] media. This autoagglutination is not necessarily damaging and divisions often initiate from such clumps. It remains to be seen whether developmentally-regulated lectins are responsible or perhaps surface charge effects.

Secondly some caution must be exercised with regard to the time between protoplast isolation and use, and the medium in which they have been stored. We have observed a dramatic effect of time in culture on agglutinability. The agglutinability with peanut lectin (PNA) or β-lactosyl Yariv antigen (β-LAC) drops dramatically in the first day or two of culture in normal medium. Coumarin[101] and 2,6-dichloro-benzonitrile (2,6-DB)[101] are both inhibitors of cell wall synthesis. In the presence of these compounds the drop in agglutinability is delayed (Fig. 6). This suggests that wall formation is responsible for the reduced agglutinability. In the case of ConA the first 20 to 30 hours in culture actually enhance agglutinability but thereafter wall formation interferes. Hanke[25] reported a drop with time in SBA-induced

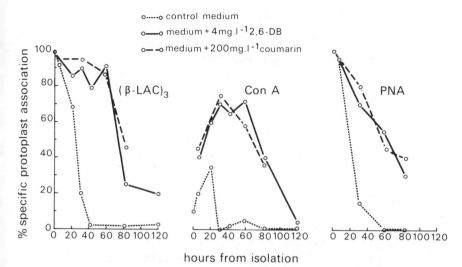

Figure 6. The effect of culture of Petunia hybrida protoplasts on their agglutinability with β-lactosyl Yariv
antigen (β-LAC), Con A and PNA. Culture was both with and
without the cell wall synthesis inhibitors coumarin or
2,6-DB, as indicated. The percent specific protoplast association was determined as the difference between the controls
(no agglutinin) and the tests each time and for each medium.

agglutination of soybean protoplasts but a rise with time in
ConA-induced agglutination.

These results with ConA introduce another issue for
caution. It is possible that the isolation procedures for
protoplasts, which involve crude enzyme preparations, remove
important recognition components or even expose artifactual
determinants. The preincubation of petunia and Nicotiana
debneyi protoplasts for 1 or 2 days in the presence of
coumarin or 2,6-DB greatly enhanced their agglutinability
with ConA (Fig. 6). This suggests that synthetic metabolism
enhances the receptors for ConA. Those receptors are unlikely to be wall polymers.

The fact that wall formation inhibits agglutination
does not necessarily negate the in vivo significance of the
interaction being investigated. Cell deformability is a

very critical factor in agglutination.[102] [103] Naked protoplasts are highly deformable. The reformation of wall will
drastically reduce deformability and consequently agglutinability even if the binder/determinant interactions are
occurring in the presence of the wall as they must in nature.

ACKNOWLEDGEMENTS

The Yariv antigens were a gift from Dr. M. A. Jermyn.
I wish to thank colleagues at the Waite Institute, Adelaide,
the Fredrich Miescher Institute, Basel, and the CSIRO
Division of Plant Industry, Canberra for helpful discussions.

REFERENCES

1. Kauss, H. 1976. Plant lectins (phytohemagglutinins).
    Prog. Bot. [Fortschr. Bot.] 38:58-70.
2. Liener, I.E. 1976. Phytohemagglutinins (phytolectins).
    Annu. Rev. Plant Physiol. 27:291-319.
3. Rüdiger, H. 1978. Lectine, pflanzliche zuckerbindende
    Proteine. Naturwissenschaften 65:239-244.
4. Nozue, M., K. Tomiyama and N. Doke. 1979. Evidence for
    adherence of host plasmalemma to infecting hyphae of
    both compatible and incompatible races of Phytophthora
    infestans. Physiol. Plant Pathol. 15:111-115.
5. Shimony, C. and J. Friend. 1976. Ultrastructure of
    the interaction between Phytophthora infestans (Mont).
    de Bary and tuber discs of potato (Solanum tuberosum
    L.) cv. King Edward. Physiol. Plant Pathol.
    11:243-249.
6. Hohl, H.R. and E. Suter. 1976. Host-parasite inter-
    faces in a resistant and a susceptible cultivar of
    Solanum tuberosum inoculated with Phytophthora infestans:
    leaf tissue. Can J. Bot. 54:1956-1970.
7. Asada, Y. and M. Shiraishi. 1979. Discontinuity of the
    plasma membrane of Raphanus sativus around haustoria of
    Peronospora parasitica. In Biochemistry and Cytology
    of Plant-Parasite Interactions. (K. Tomiyama et al.,
    eds.). Kodansha Ltd. Tokyo. pp. 32-34.
8. Ouchi, S., H. Oku and C. Hibino. 1976. Some character-
    istics of induced susceptibility and resistance demon-
    strated in powdery mildew of barley. ibid. pp. 181-184.
9. Tomiyama, K., N. Doke and H.S. Lee. 1976. Mechanisms
    of hypersensitive cell death in host-parasite inter-
    action. ibid. pp. 136-142.

10. Yeoman, M.M., D.C. Kilpatrick, M.B. Miedzybrodzka and A.R. Gould 1978. Cellular interactions during graft formation in plants, a recognition phenomenon? In Cell-Cell Recognition. Symp. 32, Soc. Exp. Biol. Cambridge University Press. Cambridge. pp. 139-160.
11. Roland, J.-C. 1973. The relationship between the plasmalemma and plant cell wall. Intern. Rev. Cytol. 36:45-92.
12. Bowles, D.J. and H. Kauss. 1975. Carbohydrate-binding proteins from cellular membranes of plant tissue. Plant Sci. Lett. 4:411-418.
13. Bowles, D.J. and H. Kauss. 1976. Characterisation, enzymatic and lectin properties of isolated membranes from Phaseolus aureus. Biochim. Biophys. Acta 443:360-374.
14. Clarke, A.E., R.B. Knox and M.A. Jermyn. 1975. Localization of lectins in legume cotyledons. J. Cell Sci. 19:157-167.
15. Bowles, D.J., H. Lis and N. Sharon. 1979. Distribution of lectins in membranes of soybean and peanut plants. I. General distribution in root, shoot and leaf tissue at different stages of growth. Planta 145:193-198.
16. Bowles, D.J. 1979. Lectins as components of plant membranes. In Plant Organelles. (E. Reid, ed.). Ellis Horwood, Chichester. pp. 165-171.
17. Kauss, H. 1976. Plant lectins (phytohemagglutinins). Prog. Bot. 38:58-70.
18. Kauss, H. and D.J. Bowles. 1976. Some properties of carbohydrate-binding proteins (lectins) solubilized from cell walls of Phaseolus aureus. Planta 130:169-174.
19. Kilpatrick, D.C., M.M. Yeoman and A.R. Gould. 1979. Tissue and subcellular distribution of the lectin from Datura stramonium (thorn apple). Biochem. J. 184:215-219.
20. Glimelius, K., A. Wallin and T. Eriksson 1974. Agglutinating effects of Concanavalin A on isolated protoplasts of Daucus carota. Physiol. Plant. 31:225-230.
21. Williamson, F.A., L.C. Fowke, F.C. Constabel and O.L. Gamborg. 1976: Labelling of Concanavalin A sites on the plasma membrane of soybean protoplasts. Protoplasma 89:305-316.
22. Burgess, J. and P.J. Linstead. 1976. Ultrastructural studies of the binding of Concanavalin A to the plasmalemma of higher plant protoplasts. Planta 130:73-79.

23. Larkin, P.J. 1978. Plant protoplast agglutination by
     lectins. Plant Physiol. 61:626-629.
24. Chin, J.C. and K.J. Scott. 1979. Effect of phyto-
     lectins on isolated protoplasts from plants. Ann. Bot.
     43:33-44.
25. Hanke, D.E. 1979. Plasma-membrane surface components
     investigated using protoplasts. In Plant Organelles.
     (Eric Reid, ed.) Chichester, Ellis Horwood Ltd. pp.
     196-198.
26. Larkin, P.J. unpublished.
27. Williamson, F.A. 1979. Concanavalin A binding sites on
     the plasma membrane of leek stem protoplasts. Planta
     144:209-215.
28. Nicolson, G.L. 1974. The interactions of lectins with
     animal cell surfaces. Intern. Rev. Cytol. 39: 89-190.
29. Lotan, R., H. Lis, A. Rosenwasser, A. Novagrodsky and
     N. Sharon. 1973. Enhancement of the biological ac-
     tivities of soybean agglutinin by cross-linking with
     glutaraldehyde. Biochem. Biophys. Res. Comm.
     55:1347-1355.
30. Hamblin, J. and S.P. Kent. 1973. Possible role of
     phytohaemagglutinin in Phaseolus vulgaris L. Nature
     (New Biol.) 245:28-30.
31. Bohlool, B.B. and E.L. Schmidt 1974. Lectins: a
     possible basis for specificity in the Rhizobium-legume
     root nodule symbiosis. Science 185:269-271.
32. Dazzo, F.B. and D.H. Hubbell. 1975. Antigenic dif-
     ferences between infective and non-infective strains of
     Rhizobium trifolii. Appl. Microbiol. 30:172-177.
33. Dazzo, F.B. and D.H. Hubbell. 1975. Cross reactive
     antigens and lectin as determinants of symbiotic spec-
     ificity in the Rhizobium-clover association. Appl.
     Microbiol. 30:1017-1033.
34. Wolpert, J.S. and P. Albersheim. 1976. Host-symbiont
     interactions I The lectins of legumes interact with the
     0-antigen-containing lipopolysaccharides of their
     symbiont Rhizobia. Biochem. Biophys. Res. Commun.
     70:729-737.
35. Dazzo, F.B., W.E. Yanke and W.J. Brill. 1978. Trifoliin:
     a Rhizobium recognition protein from white clover.
     Biochim. Biophys. Acta 539:276-286.
36. Dazzo, F.B. and W.J. Brill. 1978. Regulation by fixed
     nitrogen of lost-symbiont recognition in the Rhizobium-
     clover symbiosis. Plant Physiol. 62:18-21.

37. Bhuvaneswari, T.V. and W.D. Bauer. 1978. Role of lectin in plant-microorganism interactions III Influence of rhizosphere/rhizoplane culture conditions on the soybean lectin-binding properties of rhizobia. Plant Physiol. 62:71-74.
38. Calvert, H.E., M. Lalonde, T.V. Bhuvaneswari and W.D. Bauer. 1978. Role of lectins in plant-microorganism interactions. IV Ultrastructural localization of soybean lectin binding sites on Rhizobium japonicum. Can. J. Microbiol. 24:784-793.
39. Kato, G., Y. Maruyama and M. Nakamura. 1979. Role of lectins and lipopolysaccharides in the recognition process of specific legume-Rhizobium symbiosis. Agric. Biol. Chem. 43:1085-1092.
40. Schmidt, E.L. 1979. Initiation of plant root-microbe interactions. Ann. Rev. Microbiol. 33:355-376.
41. Ozawa, T. and M. Yamaguchi. 1979. Inhibition of soybean cell growth by the adsorption of Rhizobium japonicum. Plant Physiol. 64:65-68.
42. Law, I.J. and B.W. Strijdom. 1977. Some observations on plant lectins and Rhizobium specificity. Soil Biol. Biochem. 9:79-84.
43. Chen, A.T. and D.A. Phillips. 1976. Attachment of Rhizobium to legume roots as the basis for specific interactions. Physiol. Plant. 38:83-88.
44. Gill, D.M. and C.A. King. 1975. The mechanism of action of cholera toxin in pigeon erythrocyte lysates. J. Biol. Chem. 250:6424-6432.
45. Ledley, F.D., G. Lee, L.D. Lohn, W.H. Habig and M.C. Hardegree. 1977. Tetanus toxin interactions with thyroid plasma membranes. Implications for structure and function of tetanus toxin receptors and potential pathophysiological significance. J. Biol. Chem. 252:4049-4055.
46. Draper, R.K., D. Chin and M.I. Simon. 1978. Diphtheria toxin has the properties of a lectin. Proc. Natl. Acad. Sci. U.S.A. 75:261-265.
47. Hartmann, J.X., K.N. Kao, O.L. Gamborg and R.A. Miller. 1973. Immunological methods for the agglutination of protoplasts from cell suspension cultures of different genera. Planta 112:45-56.
48. Burgess, J. and E.N. Fleming. 1974. Ultrastructural studies of the aggregation and fusion of plant protoplasts. Planta 118:183-193.

49. Strobel, G.A. and W.H. Hess. 1974. Evidence for the
     presence of toxin-binding protein on the plasma membrane
     of sugarcane cells. Proc. Natl. Acad. Sci. U.S.A.
     71:1413-1417.
50. Larkin, P.J. 1977. Plant protoplast agglutination and
     membrane-bound β-lectins. J. Cell Sci. 26:31-46.
51. Raff, J., I.F.C. McKenzie, and A.E. Clarke. 1980.
     Antigenic determinants of Prunus avium are associated
     with the protoplast surface. Z. Pflanzenphysiol.
     98:225-234.
52. Yariv, J., M.M. Rapport and L. Graf. 1962. The inter-
     action of glycosides and saccharides with antibody to
     the corresponding phenylazo glycosides. Biochem. J.
     85:383-388.
53. Jermyn, M.A. and Y.M. Yeow. 1975. A class of lectins
     present in the tissues of seed plants. Aust. J. Plant
     Physiol. 2:501-531.
54. Larkin, P.J. 1978. Plant protoplast agglutination by
     artificial carbohydrate antigens. J. Cell Sci.
     30:283-292.
55. Gleeson, P.A. and A.E. Clarke. 1979. Structural studies
     on the major component of Gladiolus style mucilage, an
     arabinogalactan protein. Biochem. J. 181:607-621.
56. Clark, A.E., P. Gleeson, S. Harrison and R.B. Knox.
     1979. Pollen-stigma interactions: identification and
     characterization of surface components with recognition
     potential. Proc. Natl. Acad. Sci. U.S.A. 76:3358-3362.
57. Clarke, A.E., R.L. Anderson and B.A. Stone. 1979. Form
     and function of arabinogalactans and arabinogalactan-
     proteins. Phytochemistry 18:521-540.
58. Baldo, B.A., H. Neukon, B.A. Stone and G. Uhlenbruck.
     1978. Reaction of some invertebrate and plant agglu-
     tinins and a mouse myeloma anti-galactan protein with
     an arabinogalactan from wheat. Aust. J. Biol. Sci.
     31:149-160.
59. Majumdar, T. and A. Surolia. 1978. Cross-linked arabino-
     galactan - new affinity matrix for purification of
     Ricinus communis lectins. Experientia 34:979-980.
60. Majumdar, T. and A. Surolia. 1978. A large scale pre-
     paration of peanut agglutinin on a new affinity matrix.
     Prep. Biochem. 8:119-131.
61. Larkin, P.J. 1977. The use of cell surface properties
     for hybrid protoplast selection. Ph.D thesis.
     University of Adelaide.

62. Jermyn, M. 1977. β-Lectin-flavonol interactions.
     Arabinogalactan Protein News 1: 37-38. Proceedings
     of the Arabinogalactan Protein Club. University of
     Melbourne. Melbourne, Australia.
63. Peters, B.M., D.H. Cribbs and D.A. Stelzig. 1978.
     Agglutination of plant protoplasts by fungal cell wall
     glucans. Science 201:364-365.
64. Shirey, R.E. and D.A. Stelzig. 1980. Structural charac-
     terization of an elicitor enzymatically released from
     cell walls of Phytophthora infestans. Phytopathology
     (In press).
65. Tomiyama, K. 1971. Cytological and Biochemical studies
     of the hypersensitive reaction of potato cells to
     Phytophthora infestans. In Morphological and Biochemi-
     cal Events in Plant-Parasite Interaction. (S. Akai and
     S. Ouchi, eds.). The Phytochem. Soc. Japan. Tokyo.
     pp. 387-401.
66. Kuč, J., W. Currier, J. Elliston and J. McIntyre. 1976.
     Determinants of plant disease resistance and suscep-
     tibility: a perspective based on three plant-parasite
     interactions. In Biochemistry and Cytology of Plant-
     Parasite Interaction. (K. Tomiyama et al., eds.).
     Kodansha Ltd. Tokyo. pp. 168-180.
67. Marcan, H., M.C. Jarvis and J. Friend. 1979. Effect of
     methyl glycosides and oligosaccharides on cell death
     and browning of potato tuber discs induced by mycelial
     components of Phytophthora infestans. Physiol. Plant
     Pathol. 14:1-9.
68. Kim, W.K. and I. Uritani. 1974. Fungal extracts that
     induce phytoalexins in sweet potato roots. Plant Cell
     Physiol. 15:1093-1098.
69. Albersheim, P. and B.S. Valent. 1978. Host-pathogen
     interactions in plants. Plants, when exposed to oligo-
     saccharides of fungal origin, defend themselves by
     accumulating antibiotics. J. Cell Biol. 78:627-643.
70. Uritani, I. 1978. Biochemistry of host response to
     infection. In Progress in Phytochemistry. Vol. 5
     (L. Reinhold et al., eds.). Pergamon, London. pp. 29-63.
71. Keen, N.T. 1976. Specific elicitors of phytoalexin
     production: determinants of race specificity. In Bio-
     chemistry and Cytology of Plant-Parasite Interaction.
     (K. Tomiyama et al., eds.). Kodansha Ltd. Tokyo.
     pp. 84-93.

72. Keen, N.T. 1978. Surface glycoproteins of Phytophthora
     megasperma var. sojae function as race specific glyceol-
     lin elicitors in soybeans. (Abstr.) Phytopathol. News
     12:221.
73. van Dijkman, A. and A.K. Sijpesteijn. 1973. Leakage
     of preabsorbed ^{32}P from tomato leaf discs infiltrated
     with high molecular weight products of incompatible
     races of Cladosporium fulvum. Physiol. Plant Path.
     3:57-67.
74. Lazarovits, G., B.S. Bhullar, H.J. Sugiyama and V.J.
     Higgens. 1979. Purification and partial characteriza-
     tion of a glycoprotein toxin produced by Cladosporium
     fulvum. Phytopathology 69:1062-1068.
75. Doke, N., N.A. Garas and J. Kuć. 1980. Effect on host
     hypersensitivity of suppressors released during the
     germination of Phytophthora infestans cytospores.
     Phytopathology 70:35-39.
76. El-Banoby, F.E. and K. Rudolph. 1979. Induction of
     water-soaking in plant leaves by extracellular poly-
     saccharides from phytopathogenic pseudomonads and
     xanthomonads. Physiol. Plant Pathol. 15:341-349.
77. Nachmias, A., I. Barash, V. Buchner, Z. Solel and G. A.
     Strobel. 1979. A phytotoxic glycopeptide from lemon
     leaves infected with Phoma tracheiphila. Physiol.
     Plant Pathol. 14:135-140.
78. Rancillac, M., R. Kaur-Sawhney, B. Straskawicz and A.W.
     Galston. 1976. Effects of cycloheximide and kinetin
     pretreatments on responses of susceptible and resistant
     Avena leaf protoplasts to the phyto-toxin victorin.
     Plant and Cell Physiol. 17:987-995.
79. Wiese, L. and W. Wiese. 1978. Sex cell contact in
     Chlamydomonas, a model for cell recognition. In Cell-
     Cell Recognition. Symp. 32, Soc. Exp. Biol.
     pp. 83-103.
80. Lippincott, B.B., M.H. Whatley and J.A. Lippincott.
     1977. Tumor induction by Agrobacterium involves attach-
     ment of the bacterium to a site on the host plant cell
     wall. Plant Physiol. 59:388-390.
81. Lippincott, J.A. and B.B. Lippincott. 1978. Cell walls
     on crown-gall tumors and embryonic plant tissues lack
     Agrobacterium adhesion sites. Science 199:1075-1078.
82. Glogowski, W. and A.G. Galsky. 1978. Agrobacterium
     tumefaciens site attachment as a necessary pre-requisite
     for crown gall tumor formation on potato discs. Plant
     Physiol. 61:1031-1033.

83. Smith, V.A. and J. Hindley. 1978. Effect of agrocin 84 on attachment of Agrobacterium tumefaciens to cultured tobacco cells. Nature 276:498-500.
84. Ohyama, K., L.E. Pelcher, A. Schaeffer and L.C. Fowke. 1979. In vitro binding of Agrobacterium tumefaciens to plant cells from suspension culture. Plant Physiol. 63:382-387.
85. Sequeira, L., G. Gaard and G.A. de Zoeten. 1977. Interaction of bacteria and host cell walls: its relation to mechanisms of induced resistance. Physiol. Plant Pathol. 10:43-50.
86. Sequeira, L. 1978. Lectins and their role in host-pathogen specificity. Annu. Rev. Phytopathol. 16:453-481.
87. Whatley, M.H., N. Hunter, M.A. Cantrell, C. Hendrick, K. Keegstra and L. Sequeira. 1980. Lipopolysaccharide composition of the wilt pathogen, Pseudomonas solanacearum. Plant Physiol. 65:557-559.
88. Sequeira, L. and T.L. Graham. 1977. Agglutination of avirulent strains of Pseudomonas solanacearum by potato lectin. Physiol. Plant Pathol. 11:43-54.
89. Goodman, R.N., P.-Y. Huang, J.S. Huang and V. Thaipanich. 1976. Induced resistance to bacterial infection. In Biochemistry and Cytology of Plant-Parasite Interaction. (K. Tomiyama et al., eds.). Kodansha Ltd. Tokyo. pp. 35-42.
90. Horino, O. 1976. Induction of bacterial leaf blight resistance by incompatible strains of Xanthomonas oryzae in rice. ibid pp. 43-55.
91. Anderson, A.J. and C. Jasalavich. 1979. Agglutination of pseudomonad cells by plant products. Physiol Plant Pathol. 15:149-159.
92. Sing, V.O. and M.N. Schroth. 1977. Bacteria-plant cell surface interactions: active immobilization of sapro-phytic bacteria in plant leaves. Science 197:759-761.
93. Kameya, T. 1973. The effects of gelatin on aggregation of protoplasts from higher plants. Planta 115:77-82.
94. Keller, F. personal communication in ref. 57.
95. Kao, K.N. and M.R. Michayluk. 1974. A method for high-frequency intergeneric fusion of plant protoplasts. Planta 115:355-367.
96. Wallin, A., K. Glimelius and T. Eriksson. 1974. The induction of aggregation and fusion of Daucus carota protoplasts by polyethylene glycol. Z. Pflanzenphysiol. 74:64-80.

97. Nagata, T. 1978. A novel cell-fusion method of proto-
    plasts by polyvinyl alcohol. Naturwissenschaften
    65:263-264.
98. Hankins, C.N., J.I. Kindinger and L.M. Shannon. 1980.
    Legume α-galactosidases which have hemagglutinin
    properties. Plant Physiol. 65:618-622.
99. Gamborg, O.L., R.A. Miller and K. Ojima. 1968.
    Nutrient requirements of suspension cultures of soybean
    root cells. Exp. Cell. Res. 50:151-158.
100. Burgess, J. and P.J. Linstead. 1977. Coumarin inhibi-
    tion of microfibril formation at the surface of cul-
    tured protoplasts. Planta 133:267-273.
101. Meyer, Y. and W. Herth. 1978. Chemical inhibition of
    cell wall formation and cytokinesis but not nuclear
    division, in protoplasts of Nicotiana tabacum L. cul-
    tured in vitro. Planta 142:253-262.
102. Wang, P.Y., D.W. Evans and W. Zingg. 1980. Concanavalin
    A agglutinability of dextran gel spheres. A physical
    model for cell agglutinability. Biochim. Biophys. Acta
    628:228-239.
103. Gibson, G.A., M.D. Marquardt and J.A. Gordon. 1975.
    Cell rigidity: effect on concanavalin A-mediated
    agglutinability of fibroblasts after fixation. Science
    189:45-46.

Chapter Eight

MOLECULAR ASPECTS OF RECOGNITION AND RESPONSE IN THE
POLLEN-STIGMA INTERACTION

ADRIENNE E. CLARKE AND PAUL A. GLEESON

School of Botany
University of Melbourne
Parkville 3052, Australia

INTRODUCTION

Although fertilization in plants is fundamental to life,
our understanding of the process at the molecular level is
restricted to a few facets of a few systems; predictably,
the best information has come from the simplest systems.
For example, the first event of fertilization in some algal
species is direct membrane contact of gametes which do not

have cell walls, and the recognition of compatible gametes
is apparently mediated by essentially the same sort of
reaction which occurs during the mutual recognition of
animal cells. The cellular events leading to fertilization
in higher plants are more complex: in most cases not only
are there a series of interactions between the haploid pollen housing the male gametes, and the diploid female tissues of the pistil, prior to fertilization, but the interactions involve contact of cells which have walls over-
laying the plasma membrane. For this reason, we will
consider the possible mechanisms for cell-cell recognition
between plant cells. We will then outline the biology of
fertilization in flowering plants with particular reference
to the stages where cell-cell recognition interactions are
evident. The information available concerning the nature
of the surface components of cells which come into contact
during pollination will then be considered, and finally we
will present some experimental data on the structure and
function of three classes of macromolecules implicated in
the events of pollination.

MOLECULAR BASIS OF CELL-CELL RECOGNITION

      Cell recognition is the initial event of cell-cell
communication which elicits a defined biochemical, physiological or morphological response.[1] For animal cells, the
receipt and translation of intercellular signals is believed
to be largely a cell surface phenomenon; furthermore, it
depends on formation of a complex between the signalling
molecule and a membrane bound receptor. The reaction is
specific both for receipt of a particular signal and for a
particular cellular response. There are two general mechanisms envisaged for cell recognition: In the first, complex formation is at the surfaces of cells in contact, and
is mediated by one or more pairs of complementary molecules
(Fig. 1a) of which at least one is protein in nature. In
many instances the partners in this interaction are saccharide components of glycoproteins or glycolipids and complementary proteins or glycoproteins (for reviews see
references 2-5). If these saccharide-binding components
are multivalent, they can be classified as lectins.[6] For
the second major mechanism, it is assumed that cells may
have specific membrane receptors for extracellular macromolecules (Fig. 1c); if the macromolecule is multivalent,
it may effectively cross link cells (Fig. 1b). Implicit

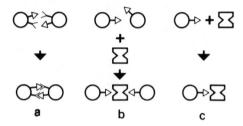

Figure 1. Possible mechanisms by which recognition in animal cells may be mediated. (a) Interaction of one or more pairs of sterically complementary molecules; (b) Cell attachment via multivalent extracellular molecule; (c) Attachment of extracellular molecule to plasma membrane receptor.

in both mechanisms is the idea that binding of a macromolecule to the surface receptor*, whether it is itself part of another cell surface or is extracellular, initiates some change in the membrane which is then transmitted to the cytoplasm, eventually to give the observed response. Various possibilities are, that movement of the membrane receptors induced by the binding may cause permeability changes in the membrane, activation of peripheral enzymes, alterations in the underlying cytoskeleton or internalization of the bound receptors which then move through the cytoplasm to the target organelle, to initiate the response (Fig. 2).[10]

---

*In this discussion, we use the term "receptor" in the sense defined by Greaves[7] and Goldstein and Hayes[8] to describe a membrane structure which binds external molecules in a highly specific way, thereby transmitting signals from the environment to the cell. An alternative nomenclature has recently been suggested by Ballou and coworkers[9] in which "cognor" signifies the active, recognizer partner of the interaction, and "cognon" denotes the passive partner, which is the site recognized.

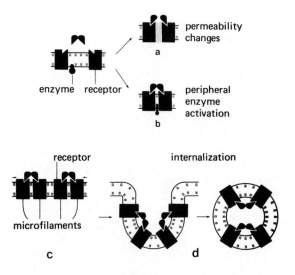

Figure 2.   Some possible mechanisms for response of a cell
to receipt of an extracellular molecular signal: Binding
of a multivalent molecule to membrane receptors may cause
(a) Permeability changes by clustering of receptors; (b)
Activation of enzymes bound to the membrane; (c) Alterations in the cytoskeletal elements; (d) Internalization
of bound macromolecules (adapted from Reference 10).

        These types of complementary interaction, (Fig. 1) in-
volving surface saccharides and complementary lectins, are
known for a number of animal cell systems.   These are also
operative in some plant cell interactions which involve
direct membrane-membrane contact.   Thus, contact of naked
gametes during mating of certain algal species depends on
binding between surface saccharides and specific surface
saccharide-binding components.   For example, mating in the
unicellular, biflagellate, fresh water alga Chlamydomonas
proceeds by adhesion at the flagella tips of gametes of the
opposite mating type, followed by cell fusion to form a
zygote.   In some species this initial adhesion possibly
involves interaction between α-mannosyl groups on the
flagellar surfaces of the plus gametes and a complementary

receptor (lectin) on the flagellar surface of the minus gametes.[11] Mating in Fucus, a member of the brown algae Fucales apparently depends on a similar saccharide-saccharide binding interaction. In this system, fertilization of the large non-motile eggs is highly species-specific and is mediated by attachment of the tip of the anterior flagellum of the sperm to the egg cell membrane. The available evidence indicates that egg cell surface saccharides containing both fucosyl and mannosyl residues are involved in the specific initiating events of fertilization.[12]

Direct communication between higher plant cells, by contact between the extracellular faces of the plasma membranes is precluded by the presence of the cell walls; however, there are two major mechanisms by which recognition of extracellular signals can be envisaged.

1. Complementary interaction of components at the wall surfaces of cells in contact

For cells which come into physical contact such as pollen on stigma surfaces, pollen tubes and the cells of the style canal, fungal zoospores at root surfaces, or fungal hyphae with epidermal cells, it is possible that direct contact between walls of cells could initiate some sort of reorganization or deformation of wall component(s), which in turn may transmit a signal to the plasma membrane and thence to the cytoplasm (Fig. 3a). One class of likely candidates for this type of interaction would be the polysaccharide matrix components, which in contrast to the fibrillar components have considerable informational potential in the type and arrangement of their constituent monosaccharides. These polysaccharides may interact with complementary sequences of polysaccharides in the interacting cell wall, with subsequent changes in the gel properties of the walls. Certainly, interactions between complementary regions of polysaccharides in solution are known,[13] and they do affect the gel properties. Alternatively, the matrix polysaccharides may be modified by interaction with proteins or glycoproteins such as lectins or enzymes. The current information regarding the molecular architecture of the cell wall is not sufficiently detailed to allow assessment of the likelihood of this possibility: we know a great deal about the structure and form of some of the wall polysaccharides and glycoproteins of a few cell types, but very

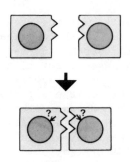

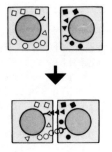

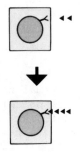

a Direct cell wall-cell wall contact. Interaction via complementary macromolecules at the surface

b Extracellular molecule interacts with complementary wall molecules at the surface

c Direct cell wall-cell wall contact. Interaction via wall macromolecules which diffuse through the wall of the interacting cell to a plasma membrane receptor

d Extracellular molecule diffuses through cell wall to a plasma membrane receptor

Figure 3. Possible mechanisms by which recognition in plant cells may be mediated. In each case a signal ultimately reaches the plasma membrane through the cell wall. In (a) and (b) the wall itself is the site of initial receipt of information - that is, the wall plays an active part in the recognition reaction. In (c) and (d) the wall is regarded as a passive barrier through which the signal molecule may pass to a membrane receptor. The square outline represents the outer boundary of the cell wall. The circle represents the plasma membrane.

little about their precise intermolecular associations and organization within the wall.[14] For instance, there may be concentration gradients of particular components, some, for example being localized at the outer wall surface rather than the region adjoining the plasma membrane. In spite of the lack of knowledge in this area, there is evidence that wall-wall contact does operate in one situation - the sex-specific agglutination of yeasts. For several species, including Hansenula wingei, specific agglutination is apparently mediated by complex formation between complementary surface molecules which are evenly distributed over the whole cell wall: furthermore the interaction involves a glycoprotein and saccharide residues (Fig. 3a).[9] A variant of this mechanism is one in which an extra-cellular macromolecular signal may be received at the outer surface of the wall and be transmitted to the plasma membrane by some mechanism involving a similar perturbation of the wall structure or secondary messenger in the wall (Fig. 3b). In this type of interaction the wall components would play an active role.

2. Transmission of an extracellular signal through the cell wall to a plasma membrane receptor

An alternative mechanism in which the wall components play a passive role can also be envisaged. It may be that a molecular signal originating from one cell, could diffuse to the cell surface and then directly through the walls of the interacting cell to be received at the plasma membrane by a complementary receptor (Fig. 3c). For cells within the same tissue this would also involve passage through the middle lamella. There is evidence that molecules of up to 5nm in diameter (equivalent to approx. MW 17,000 for a globular protein or 6,500 for a dextran) can pass through the gel matrix of the primary wall of some cells,[15] so that information transmission by this mechanism is possible. The variant of the mechanism corresponding to that shown in Fig. 3b, is that in which an extracellular macromolecular signal diffuses through the cell wall of the target cell to the plasma membrane receptor (Fig. 3d).

In spite of the lack of information regarding details of the mechanism, there is evidence that this type of inter-action may be the basis of at least one example of cell-cell recognition. This is the interaction between Phytophthora

megasperma var. sojae and soybean in which an extracellular
3,6 β-glucan, related to the fungal cell wall matrix components, will initiate phytoalexin production in both tissue
slices and cultured callus cells.[16] There is also indirect
evidence for the presence of a specific receptor for these
and related (1→3)-linked β-glucans at protoplast surfaces.[17]
In vivo the interaction could be envisaged in terms of
diffusion of a low MW fungal wall polysaccharide, through
the host plant cell wall to a plasma membrane receptor.
Binding at the membrane receptor then somehow triggers
phytoalexin biosynthesis (Fig. 2).

There are also a number of interactions between higher
plants and microorganisms in which recognition, either in a
symbiotic or pathogenic relationship probably involves surface saccharides and saccharide receptors (for reviews see
References 1, 16, 18-23).

Apart from contact via the walls, cells in somatic
tissues may also be in contact via plasmodesmata, which are
continuities of the plasma membrane through the cell wall.
Movement within these intercellular channels is apparently
restricted to molecules of up to 1,000 daltons. Communication of this kind is important in maintaining a state of
differentiation in a tissue; it allows some exchange of
intercellular information, within the symplast, and is
separate from the extracellular recognition systems discussed above, which operate within the apoplast.[24]

We will now outline the biology of fertilization in
flowering plants as a basis for considering how these views
of cell-cell recognition apply to pollination.

BIOLOGY OF FERTILIZATION IN FLOWERING PLANTS

The interacting partners are the pollen and the pistil.
Pollen grains containing the male gametes are transported,
usually by wind currents or by animal vectors to the female
organ, the pistil. If the mating is compatible, the pollen
hydrates at the stigma surface of the pistil, produces a
pollen tube which penetrates the stigma surface and grows
intercellularly through the style to the embryo sac, where
fertilization occurs (Fig. 4). This involves release of
two sperm cells, one fertilizes the egg and the other fustes
with primary endosperm nucleus. Breeding is controlled by

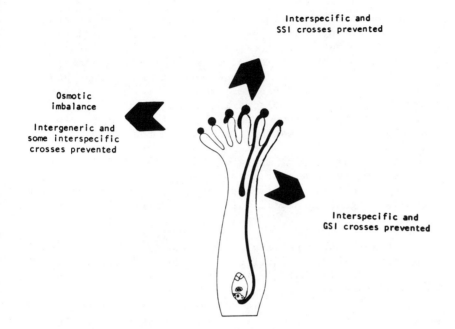

Figure 4. Diagrammatic representation of the events of
pollination and the various stages at which growth of incompatible pollen tubes may be arrested. The hand-like
structure represents the pistil, and the finger-like processes represent individual papillar cells. The solid black
structures represent the pollen grains and tubes. (Adapted
from Reference 1). (SSl = sporophytic self-incompatibility;
GSl = gametophytic self-incompatibility).

selection between the different types of pollen which may be
received by the stigma: intergeneric crosses are usually
prevented; interspecific crosses are occasionally successful, but usually intraspecific crosses are successful. The
exceptions to this are in the 70 or so plant families in
which specific self-incompatibility genes operate to prevent
inbreeding. There are usually no morphological differences
between the breeding groups, and the system favours out-
crossing by preventing self pollination.

By defining the points at which these different types
of crosses are prevented, we can start to ask questions
regarding the nature of the interacting surfaces involved
in the recognition. It is not easy to categorize these
interactions, because of the enormous variation in behavior
of foreign pollen on a pistil. However, for many intergeneric pollinations and some interspecific pollinations,
it is believed that some type of osmotic imbalance prevents
the pollen from germinating, as although the pollen may not
germinate on a particular stigma, it can be induced to
germinate in artificial media of appropriate concentration.
This barrier to germination is not necessarily based on
interaction between a cell surface receptor and an external
molecular signal, and will not be considered further in
this paper.

For some interspecific crosses and for one of the two
types of self-incompatibility (sporophytic incompatibility)
the foreign pollen may germinate, but fail to penetrate the
stigma surface. There is some information regarding the
nature of the interaction in sporophytic self-incompatibility systems of the Cruciferae and Compositae. In these
systems, germination of the pollen is controlled not by its
own genotype, but by that of the plant which produced it.
There is a single gene locus with multiple alleles; where
there is a match between an allele of the pollen parent and
the female tissues of the stigma and style, the pollen tubes
are inhibited on the stigma surface (Fig. 5a). The interaction can be viewed as a cell-cell recognition event in
which components of the stigma surface and the surface of
either the pollen grain or rudimentary pollen tube surface
interact to signal either continued or arrested growth of
the pollen tube. A most useful observation has been that
arrest of tube growth in incompatible matings is commonly
accompanied by rapid deposition of material which gives a
brilliant yellow fluorescence with aniline blue, both at the
stigma surface adjacent to the contact point with the rudimentary pollen tube and also in the tip of the tube.[25,26,27]
Because of this characteristic staining with aniline blue
the material is referred to as "callose" (see section beginning on page 194). In compatible pollinations callose is
also detected: in this situation it is laid down in pads
which apparently "block off" older sections of the tube,
restricting the cytoplasm containing the sperm cells to the
growing tip.

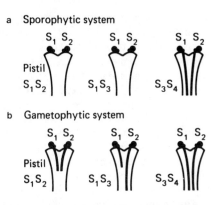

a   Sporophytic system

b   Gametophytic system

Figure 5.   Behaviour of pollen in the two major self-incompatibility systems.

(a) <u>Sporophytic</u> <u>incompatibility</u>. The pollen parent genotype is S_1S_2. When an allele in the pollen parent is matched with that of the pistil (e.g. S_1S_2 or S_1S_3) pollen germination is arrested at the stigma surface. Where there is no match (S_3S_4) the pollen may germinate and grow through the style to the embryo sac. The central diagram of Fig. 5a only applies if the allele S_1 is dominant to or codominant with S_2 in the pollen and S_1 is dominant to or codominant with S_3 in the style. If S_3 is dominant to S_1 in the style, or S_2 dominant to S_1 in the pollen, then pollen from S_2S_2 parent will be compatible. (b) <u>Gametophytic</u> <u>incompatibility</u>. The pollen parent genotype is S_1S_2. When an allele in the individual haploid pollen grain is matched with that of the diploid stylar tissues, growth of the pollen tube will be arrested, usually in the style. For example, both S_1 and S_2 pollen will be inhibited in an S_1S_2 style, but S_2 pollen will grow successfully through an S_1S_3 style. Where there is no match the pollen tube will grow through the style to the embryo sac. For example, S_1S_2 pollen grains, on an S_3S_4 pistil.

With regard to the interacting surfaces, integrity of the stigma surface is important, because by disrupting the stigma chemically (organic solvents, e.g. hexane), physically (heat, mechanical or electrical damage) or enzymically (with trypsin), pollen tubes which would normally be arrested

at the stigma surface, maybe allowed to grow into the style
(for reviews see References 1, 28, 29). There is no information available regarding the structure of the stigma surface components which are involved in this reaction. However, some information regarding the nature of the pollen
components has been derived from experiments in which
observations of "callose" production have been used as a
qualitative bioassay. For several systems, "callose" deposition can be induced at the stigma surface by application of a protein-containing fraction derived from the outer
wall (exine) of ungerminated incompatible pollen.[30,31] This
implies that exine-derived pollen proteins originating from
the anther tapetum of the pollen parent, contain information
related to the S-genotype, which is perceived at the stigma
surface. Presumably this "callose"-eliciting component is
carried onto the growing pollen tube surface, and recognition at the stigma surface results in "callose" deposition.
The finding that stigma extracts of an incompatible genotype, but not of a compatible genotype, will inhibit pollen
tube growth in vitro, indicates that the pollen tube has
some means of perceiving material of the stigma surface
and responding to this by arrest of growth (Fig. 6).

The nature of these interacting components is not
known. We would expect to see products of the S-alleles
present at the stigma surface and in the parentally derived
exine components of the pollen; attempts to demonstrate this
by immunological methods for Brassica oleracea have shown
that an antigen which correlates with the S-genotype is
indeed present in stigmas,[32,33] although no corresponding
antigen has been reported for pollen extracts.

The next point where mismatch between pollen and stigma
leads to a cut-off in the events of fertilization, is within
the style canal (Fig. 4). After some interspecific pollinations and self-pollinations in the gametophytic incompatibility system growth of pollen tubes is arrested in the
style. In gametophytic incompatibility pollen behaviour is
controlled by its own genotype. Usually there is a single
gene locus - the S locus with multiple alleles. A pollen
tube carrying a single given allele is inhibited if the
same allele is present in the style (Fig. 5b). If the
S-allele of the pollen tube is not present in the style,
then pollen tube growth proceeds through style to the
embryo sac.

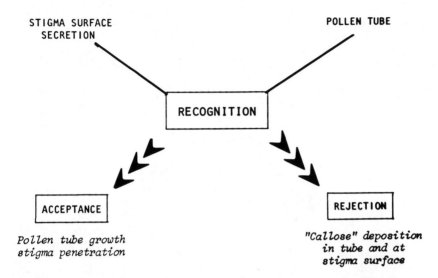

Figure 6. Diagrammatic representation of recognition and response during pollination in a sporophytic self-incompatible system.

        In most gametophytic incompatibility systems, incompatible and compatible tubes appear identical initially, but at some stage during their growth through the style, incompatible tube tips appears abnormal; in some cases the tube bursts and in other cases growth ceases and the tip becomes occluded with aniline blue-staining "callose".[34,35] In these systems, the pollen tubes grown through the style, either intercellularly through the transmitting tissue or through a hollow style canal which is filled with a mucilage at maturity. In both cases, the contact is with extracellular materials of the style. Thus again, the difference in genotype of the pollen tube and style canal is somehow perceived and the observable results are deposition of "callose" and cessation of tube growth (Fig. 7). In these cases, the interacting surfaces are that of the pollen tube and the secretions of the style canal; a pollen tube receptor for an extracellular component of the style is implied; on the basis of the previous discussion we might expect this receptor to be located either at the pollen tube surface (Fig. 3b) or at the plasma membrane (Fig. 3d). The possible

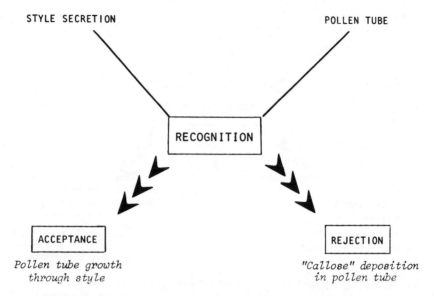

Figure 7. Diagrammatic representation of recognition and response during pollination in a gametophytic self-incompatible system.

nature of the style component involved in gametophytic self-incompatibility in Prunus avium is discussed in the section beginning on page 188. More detailed descriptions of the biology of pollination are given in references.[36,38]

NATURE OF CELL SURFACES INVOLVED IN POLLEN-STIGMA RECOGNITION

The nature of the molecules involved in the specific interactions between pllen tube ans stigma or style has not been defined for any pollination system. However, many stigmas have been examined for particular components; often the examination has been cytochemical so that the presence of certain classes of components is inferred from their staining reactions. These studies are fragmentary and it is difficult to piece together the analyses to get an overview of the compsition. For example, low molecular weight carbohydrates such as glucose, fructose and sucrose, phenolic compounds such as flavone glycosides and sterols have been detected. In many genera, stigma exudates have lipid esters of phenolic compounds as the major component; also many stigma surface secretions are enzymically active and hydrolyze artificial substrates such as α-naphthyl acetate and α-napthyl phosphate in cytochemical tests for esterase and phosphatase activity. There is even less information available regarding the composition of the style mucilage although in a pioneering study, Loewus and Labarca published a complete monosaccharide analysis of the stigmatic exudate of Lilium longiflorum[39] (see the next section).

For pollen, again, most information has been obtained cytochemically; usually the object has been either to examine the pollen grains with specific stains or to examine the wall layers for specific components. Briefly, these studies have shown that most types of pollen grains have a lipid surface coat containing carotenoid or flavanoid pigments. Some wall components diffuse into the medium when pollen is moistened; these components are derived from both the exine (outer-wall) and the intine (inner-wall). Pollen diffusates contain enzymes such as glycosidases, esterases, and glycosyl transferases. The "esterase" activity detected cytochemically has been equated with "cutinase" which is the enzyme(s) responsible for the degradation of the stigmatic cuticle by pollen.[36] Allergenic components have also been detected in grass pollen and ragweed pollen. This information is reviewed in References 1, 18, 37, 40.

Another approach to examining the nature of both pollen and stigma components has been to raise antisera to extracts and to examine the number and identity of the antigenic components. This approach was introduced by Lewis in 1952[41] and has most often been used to study S-gene products in self-incompatible pollination systems, and also to study pollen allergens. The value and limitations of this approach are discussed in the section beginning on page 188.

As to the pollen tubes there is very little analytical data or studies to indicate whether pollen tubes do secrete material, although pollen tubes have been used most sucessfully as model systems for studying cell wall growth and development.[42]

DEFINED COMPONENTS INVOLVED IN POLLINATION.

An analysis of Gladiolus stigma surface secretion and
style canal mucilage. Arabinogalactans are major components.

    Because of fragmentary information available on compo-
sition of stigma surface secretions and style mucilages and
because of their apparent role in the pollination process,
we have undertaken a detailed analysis of these secretions.
Preliminary experiments were with Gladiolus; this monocot
has large flowers with large accessible pistils and anthers,
and has the great advantage of being available commercially
throughout the year. Ideally we would have worked with defined genotypes in a self-incompatible system. However,
there are problems in collecting sufficient material for
analysis in all the available defined self-incompatible
systems - either the flowers can be grown in glasshouse
conditions, but have tiny stigmas, such as Brassica, and
the grasses, or they can be grown in the field but have a
very short flowering time, for example Prunus avium, the
sweet cherry.

    Gladiolus is self-compatible, but inter-specific
crosses arrested either at the stigma surface and/or within
the style canal, so that the pistils have capacity for
recognition.[28,43] As the stigmas of these flowers become
receptive to compatible pollen, they develop and adhesive
outer layer[44,45] which is the first point of contact with
pollen.

    Detection of arabinogalactans. This surface secretion
contains protein, carbohydrate and lipid in the ratio
20:23:0.1. Examination by SDS-polyacrylamide gel electrophoresis showed more than 15 bands which stained with
Coomassie blue, 9 of which also stained for carbohydrate
with periodate-Schiff reagent.[43] By far the most dominant
component is a high molecular weight polymer which is detected by staining the polyacrylamide gel with the
β-glucosyl artificial carbohydrate antigen.[43]

    The use of this artificial carbohydrate antigen as a
stain is important to the eventual finding that arabinogalactans are components of the stigma surface and style
canal of many plants, and in some they are major components.
The background is that in 1962, Yariv, during attempts to

detect low concentrations of antibodies to glycosides; prepared artificial carbohydrate antigens by coupling diazotized 4-aminophenyl glycosides to phloroglucinol.

These artificial antigens gave a precipitin reaction with the corresponding anti-glycosyl antibodies.[46] Later, in 1967, Yariv and his co-workers at the Weizmann Institute in Israel, made the chance observation that the β-glucosyl artifical carbohydrate antigen precipitated an arabinose- and galactose-containing polymer from soybean, jackbean and maize extracts.[47] This observation remained a curiosity for the next 8 years until Jermyn extended Yariv's observation and showed that the β-glucosyl artificial carbohydrate antigen specifically precipitated arabinogalactan-proteins from a remarkably wide range of plant seed and tissue extracts.[48,49] These artificial carbohydrate antigens are a brilliant red color which has made them valuable cytochemical stains. Usually, the β-glucosyl artificial antigen, being the most water soluble is used in the test; control experiments are performed with the α-galactosyl artificial antigen, which does not precipitate arabinogalactans.

The precise nature of the interaction between the β-glucosyl artificial carbohydrate antigen and the arabinogalactan-protein is still unresolved. However, there is some information available relating to the requirements of both the artificial carbohydrate antigen and the arabinogalactan-protein for effective interaction. In summary: analysis of the stereochemical requirements of the artificial carbohydrate antigens show that they must bear a glycopyranose residue with a β-$\underline{D}$-configuration at C(0)1 and the $\underline{D}$-gluco configuration at C(0)2.[50] A further requirement is the 1:4 orientation of the azo and glycosyloxy groups to the phenyl ring (Fig. 8). The nature of the surfaces of the arabinogalactan-protein which bind the artificial antigen is not clear, although the interaction appears likely to depend on the overall physical and chemical properties of the arabinogalactan-protein, rather than a specific binding site. The arabinogalactan-proteins which interact have similar physical and chemical properties. They are very high molecular weight polymers containing a highly branched (1→3):(1→6) β-galactan with arabinose residues in terminal positions.[51] The protein component usually represents between 5 to 15% of the molecule and is very resistant to proteolysis, indicating that the protein is

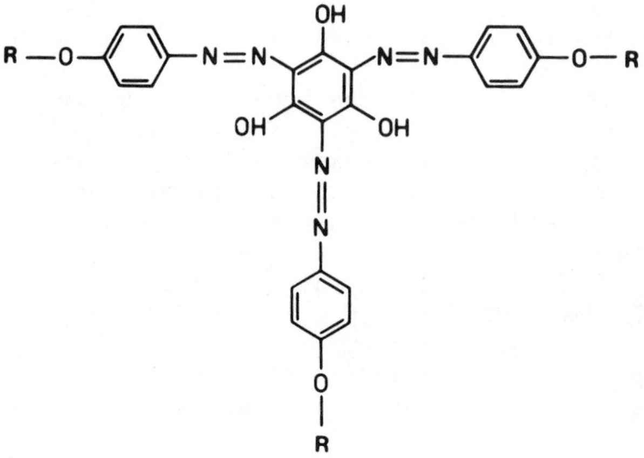

Figure 8. Structure of the β-glucosyl artificial carbohydrate antigen (R = glucosyl residue).

buried within the molecule.[52] Enzymic removal of the terminal arabinosyl residues has no apparent effect on the capacity of the molecule to interact with the artificial carbohydrate antigen (Gleeson and Clarke, unpublished observation). However, the integrity of the arabinogalactan-protein
is required for effective interaction, as mild acid hydrolysis with 12.5 mM oxalic acid abolishes its capacity for
interaction. Analysis of the arabinogalactan-protein after
this hydrolysis indicates that the galactan chains are extensively fragmented.[53] Moreover, recent experiments have
shown that the ability of the arabinogalactan-protein to
interact with the artificial antigens is destroyed by treatment with a crude galactanase preparation (Jermyn, unpublished observations) indicating that the branched galactan
portion of the molecule is somehow involved in interaction;
a more precise definition of the interacting surface is not
available at present. The most cogent evidence for the
specificity is, that of all the polysaccharides present in
plant extracts, it is always the arabinogalactans which are
precipitated on addition of a solution of a β-glucosyl
artificial carbohydrate antigen.

This β-glucosyl artificial carbohydrate antigen was used to identify the major high molecular weight component of Gladiolus stigma extracts, as an arabinogalactan. The nature stigma of Gladiolus[49] stained with this reagent and conditions for washing this material off the stigma surface while leaving the underlying cell intact were established. Sufficient surface material for analysis was obtained from 3,500 stigmas. The contents of the style canal also stained intensely with the β-glucosyl artificial carbohydrate antigen[53] (Fig. 9a). It is worth noting that the success of this stain as a cytochemical reagent, apart from its brilliant red color, also depends on its ability to precipitate the material with which it interacts; that is, it partially fixes the soluble polysaccharide during staining. This is important as it is often difficult to detect soluble cell surface carbohydrates microscopically as they are not fixed by the usual aldehyde fixatives and are often lost in the tissue preparation.[54]

Isolation of arabinogalactans. The material precipitated from style extracts by the artificial antigen contained galactose and arabinose as the major monosaccharides.[53] One drawback of isolating the polymer in this way is the difficulty of dissociation of dye from arabinogalactan; direct analysis of the insoluble complex is not satisfactory as glucose, originating from the precipitating artificial carbohydrate antigen, is always detected. An alternative procedure for isolation of the arabinogalactan which obviates this difficulty is based on affinity chromatography using an insolubilized galactosyl-binding lectin from the giant clam Tridacna maxima.[55] The binding of saccharides to this lectin depends on the presence of Ca^{++} ions, so that material bound in the presence of Ca^{++} can be simply eluted with a Ca^{++}-free buffer. In this way, the major component from both the stigma surface and style of Gladiolus was isolated.[53,56] The recovered material represented 30 to 50% of the total dry weight of the material washed from the stigma surface, (less than 1 µg per flower) and 40% of the total dry weight of the soluble style material (greater than 100 µg per flower). The material can be visualized at the stigma surface by its binding to FITC-tridacnin (Figure 9b).

Analysis of arabinogalactans. Both the stigma and style material isolated in this way behaved as single

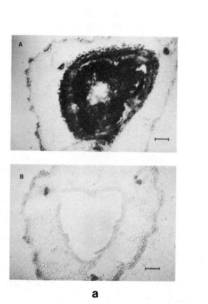

a   b

Figure 9. (a) A transverse section of the Gladiolus style
stained for the arabinogalactan-protein with β-glucosyl
artificial carbodydrate antigen (A). As a control of the
staining specificity a section was treated with α-galactosyl
artificial carbohydrate antigen (B). The bar represents
150 μM. (Adapted from Reference 48).

(b) Binding of fluoroscein isothiocyanate-tridacnin
to surface of stigmatic papillae of Gladiolus. A. Appearance of untreated papillae by scanning electron microscopy.
B. Fluorescence of surface after treatment with labelled
tridacnin. C. Absence of fluorescence after treatment with
labelled tridacnin in presence of 0.1 M lactose. Bar =
100 μM. (From Reference 44).

components on cellulose acetate electrophoresis at pH 8.8;
for each sample only one positively charged diffuse band
was detected by staining the strip with the β-glucosyl
artificial carbohydrate antigen. However, ultracentrifugal
analysis of the style material indicated that it was poly-
disperse in the molecular weight range 150,000 to 400,000.
Nevertheless, methylation analysis of fractions collected
across this molecular weight range showed that the arabinogalactan was homogeneous with respect to both monosaccharide
composition and linkage type.

Monosaccharide and methylation analyses of the stigma
and style arabinogalactans are shown in Tables 1 and 2.
The analyses are similar: both contain galactose and
arabinose as the major monosaccharides in similar proportions. The linkage composition of the two samples is also
similar: all the arabinose is present as terminal arabino-
furanosyl residues; glucose is also present solely in terminal positions. The galactose is also present solely in
terminal positions. The galactose is mainly 1,3,6-linked
with smaller amounts of (1→3)-linked, (1→6)-linked and
terminal residues. Thus, the molecules are highly

Table 1.   Monosaccharide composition of stigma and style
arabinogalactans from Gladiolus and Lilium.

MONOSACCHARIDE	COMPOSITION (% by weight)		
	Gladiolus Stigma	Gladiolus Style	Lilium Stigma* exudate
Galactose	76.1	85.8	64.1
Arabinose	20.0	14.2	30.1
Glucose	4.0	—	—
Rhamnose	—	—	5.8
Galactose: arabinose	3.8	6.0	2.1

*No analysis for uronic acids was undertaken.

Table 2.   Methylation analysis of stigma and style arabino-
            galactans from Gladiolus ad Lilium.

LINKAGE TYPE	LINKAGE COMPOSITION (moles %)		
	Gladiolus Stigma	*Gladiolus* Style	*Lilium* Stigma exudate*
Terminal rhamnosyl	0	0	7
Terminal arabinosyl	17	13	32
Terminal glucosyl	7	trace	0
Terminal galactosyl	16	29	11
1,3 linked galactosyl	13	14	10
1,6 linked galactosyl	6	6	5
1,3,6 linked galactosyl	41	39	30

* Data of Aspinall and Rosell, recalculated, excluding glucuronic acid

branched. The two molecules differ in the terminal residues,
the stigma arabinogalactan having a higher proportion of
terminal arabinose and glucose than the style arabinogalactan, which has a higher proportion of terminal
galactose.

Enzymic hydrolysis of the style arabinogalactan-protein with α-L-arabinofuranosidase confirmed the exclusive
terminal position of the arabinose residues and established
the α-configuration of these linkages. Lectin binding and
optical rotation studies indicate that the galactosyl residues are most likely linked in the β-configuration. These
data are consistent with a structure having a (1→3)-linked
β-galactan backbone with side branches of 1,6-linked
β-galactosyl residues, some of which carry the terminal
L-arabinofuranosyl residues (Fig. 10).

Also included in Table 2 are figures calculated from
the data of Aspinall and Rosell[57] on the composition of a
fraction of the surface exudate of stigmas of another mono-
cotyledon Lilium longiflorum (var. Ace) from a different

family (Liliaceae). The number of molecular species in this
fraction was not established, although the methylation
analyses indicate that a major component is an arabino-3,6-
galactan. We have confirmed this finding; freshly collected
stigma exudate from Lilium longiflorum formed a single
precipitin band with the β-glucosyl artificial carbohydrate
antigen, and the precipitated material represented about
35% dry weight of the total stigma exudate. It contained
galactose, arabinose, and rhamnose in the proportions 11:5:1
(Table 1) - a neutral sugar analysis similar to that given
by Aspinall and Rosell for the stigma exudate fraction
which they examined. Methylation analysis has shown that
although the total content of non-terminal galactosyl residues is somewhat lower in the Lilium than in the Gladiolus
arabinogalactans, the molar ratios are approximately the
same (2:1:6.6) for (1→3)- , (1→6)- and 1,3,6-linked residues. The Lilium arabinogalactan has a higher proportion of arabinose (32%) and a significant content of
terminal rhamnose residues (7%), a monosaccharide which
was not detected in the Gladiolus arabinogalactans. It
also contains a lower proportion of terminal galactosyl
residues, no terminal glucose and glucuronic acid (11%)
both as terminal units and as (1→4)-linked residues. In
contrast, the uronic acid content of the arabinogalactan of
the Gladiolus style is low (0.9%). A total monosaccharide
analysis of the high molecular weight fraction of the stigmatic exudate of Aptenia cordifolia (Aizoaceae) shows that
it contains high proportions of both galacturonic and glucuronic acids as well as a range of neutral monosaccharides
(Glu, Fru, Gal, Ara, Man, Xyl and Rha).[58]   In this respect
it differs from the Gladiolus and Lilium stigmatic exudates.

Because the Gladiolus style arabinogalactan was available in larger amounts, we have been able to characterize
it in more detail.[53]   This arabinogalactan is associated
with 3% protein; as the protein remains associated with the
carbohydrate after chromatography in 8 M urea, it is likely
to be covalently bound, although the carbohydrate-protein
linkage point(s) has not been identified. Whether protein
is associated with the stigma arabinogalactans is not known.

Characterization of the Gladiolus style arabinogalactan-
protein also provides evidence in favour of the backbone
type structure shown in Fig. 10, rather than other possible
models such as a branch-on-branch type structure or one

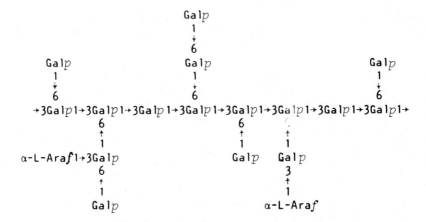

Figure 10. A proposed structure for a portion of the
Gladiolus style arabinogalactan-protein.

involving mixed (1→3);(1→6) linkage galactan chains. This
additional evidence comes, first of all, from observations
of the solubility of the molecule, before and after modification. The arabinogalactan-protein is extremely water
soluble (> 5 mg/ml); this solubility may reflect the highly
branched nature of the molecule. Linear homopolymers containing (1→3) β-linkages (e.g. paramylon) on the other hand,
are often insoluble as individual polysaccharide chains can
aggregate in a highly ordered fashion.[59] Therefore, it
might be predicted that modification of the side chain
character of this molecule to facilitate interaction between
(1→3) β-galactan backbones would result in a decrease in
solubility. Indeed, enzymic removal of terminal arabinosyl
residues does reduce the solubility of the molecule from
more than 5 mg/ml to 1 mg/ml.[53] Another modification of
the side branches which decreases the solubility is periodate oxidation. On the basis of the structure shown in
Fig. 10, periodate oxidation would be expected to cause
extensive degradation of the side branches containing
(1→6)-linked galactosyl residues as well as the terminal

galactosyl and arabinosyl residues, but no oxidation of
the (1→3)-linked galactan backbone. The oxidized arabinogalactan-protein remained in solution when stored at 4°C;
however, after freeze-drying or freezing the solution at
-20°C, it became quite insoluble.[60] These marked changes
in solubility are compatible with a backbone-type structure,
but are incompatible with a branch-on-branch type structure.

Immunological analysis of arabinogalactans. Supporting
evidence for the backbone model was also obtained from
immunochemical studies.[60] These studies were undertaken
primarily to provide a characterized specific antiserum as
an additional structural and cytochemical probe for the
investigation of these polymers in other tissues. An antiserum was raised in rabbits to the isolated style arabinogalactan-protein. The specificity of the antiserum was
investigated by immunoprecipitation using [³H]arabino-
galactan-protein. The [³H]-label was introduced into the
arabinogalactan-protein by oxidation of the terminal
galactosyl residues with galactose oxidase followed by reduction with sodium [³H]borohydride.

The hapten concentrations required for 50% inhibition
of the binding between [³H]-arabinogalactan-protein and
antiserum are shown in Table 3. The specificity is directed
primarily to the monosaccharide components, D-galactose and
L-arabinose. The disaccharide 6-0-β-D-galactopyranosyl-
galactopyranose was the most potent inhibitor and the antiserum showed preference for galactosides in the β-configuration.

The antigenic features of the arabinogalactan-protein
were investigated by examining the interaction of the antiserum with chemically and enzymically modified arabinogalactan-protein.[60] The most significant finding was that
periodate oxidation destroyed all the antigenic determinants
as the oxidized material was a poor inhibitor of the homologous interaction.

In summary, the antiserum to the Gladiolus style
arabinogalactan-protein is directed to the monosaccharides
arabinose and galactose present in the accessible side
chains while the inaccessible (1→3)-galactosyl residues of
the backbone are not apparently involved.

Table 3.   Comparison of saccharides as hapten inhibitors
of the [³H]arabinogalactan-protein-specific antiserum binding.

Inhibitor	Concentration required for 50% inhibition (mM)
6-0-β-D-Galactopyranosyl-	
D-galactopyranose	0.08
Methyl β-D-galactopyranoside	6.6
3-0-β-D-Galactopyranosyl-	
D-arabinofuranose	7.6
4-0-β-D-Galactopyranosyl-	
D-glucopyranose (lactose)	12.5
D-Galactose	23
L-Arabinose	30
Methyl α-D-galactopyranoside	100

The cross-reactivity of various polysaccharides and
glycoproteins with this antiserum was examined using an
assay based on coupling the antiserum to red blood cells
and examining the ability of the polysaccharides to agglutinate these antibody-coated red cells. Haemagglutination
with the native Gladiolus style arabinogalactan-protein was
detectable at 31 µg/ml. The Galdiolus stigma arabinogalactan-protein also cross-reacted strongly as did the
preparation from Lilium longiflorum stigma exudate. In
general, the more distant the structural relationship,
the less cross-reactivity with the specifid antiserum.[60]

Overall, these investigations of the Gladiolus and
Lilium arabinogalactans show that they are major components
of the stigma and style secretions and that they have

similar core structures with differences being expressed in the side branch monosaccharides.

The next question is, whether these arabinogalactans are present in the female sexual tissues of all plants. To answer this question, a number of style extracts have been screened for their capacity to interact with the β-glucosyl artificial carbohydrate antigen. Positive reactions were found for Prunus avium, Primula spp., Petunia spp. and Secale cereale, indicating the presence of arabinogalactans. We have now started to use the defined antiserum to Galdiolus style arabinogalactan-protein to confirm the presence of arabinogalactans in female sexual tissues of these and other species.

Structurally, these arabinogalactans are part of a widely distributed group of macromolecules - the arabino-galactans and arabinogalactan proteins, which all have similar structural features; macromolecules of this group are secreted into the medium of many callus cultures, are a major component of many plant gums and are found in extracts of a wide range of plant tissues (for review see Reference 51). In addition, they are apparently present at the sur-face of protoplasts in a number of different species.[61] It is also of interest that the arabinogalactans of the Gladiolus sexual tissues are apparently tissue specific as they differ from the arabinogalactans of the somatic tissues of Gladiolus in their monosaccharide composition, electro-phoretic behaviour and reactivity towards various galactose-binding macromolecules.[62]

The question of the role of these arabinogalactans in the pollination process remains unresolved: a number of functions consistent with their physical and chemical pro-perties are possible. Firstly, they may hold the key to specificity in the pollen-stigma interaction; this seems unlikely in view of the wide distribution of this class of macromolecules, although variations in the terminal saccharide sequences could contain specific information. Indeed, the terminal sequences in the arabinogalactans of the genus Acacia have been considered as a possible taxo-nomic marker.[63] Secondly, arabinogalactans at the stigma surface could act as adhesive gums, binding pollen grains to the surface. The Gladious stigma surface is adhesive and effectively binds a number of macromolecules.[43] Also,

the highly branched nature of the arabinogalactan fulfils
the requirements of an adhesive base; but there is no experimental evidence showing that the isolated arabinogalactans
are effective adhesives. Thirdly, the arabinogalactans may
play a nutritional role, acting as a source of carbohydrate
to nurture pollen tube growth. In Lilium longiflorum,
Labarca and Loewus[64, 65] showed that growing pollen tubes
incorporate exogenous substrate into their walls: when
high molecular weight labelled stigma exudate was injected
directly into the style canal, at least 25% of the carbohydrate of the excised pollen tubes was shown to be derived
from the labelled exudate. As an arabinogalactan is a major
component of the Lilium exudate, it seems likely that it may
play a part in nutrition of the pollen tubes. Finally, the
arabinogalactans may play a passive role in forming a matrix
which supports pollen germination and pollen tube growth.

These studies have given us detailed information regarding the structure of the major component of the stigma
and style secretions of Gladiolus: however, the structure
and function of the range of other components detected in
these secretions[43] have not been examined.

An immunological analysis of the style mucilage of
Prunus avium, an antigenic component which correlates with
S-genotype can be identified.

The problem of which component(s) is involved in the
expression of specificity in pollination has been approached
by examining a gametophytic self-incompatibility system -
Prunus avium - the sweet cherry, by both immunological and
traditional analytical methods.

As macromolecules from plant cells, like those from
animal cells are often antigenic; antibodies to these molecules can be raised and are useful tools in identification
of their structure and function. In 1952 Lewis[41] pioneered
the application of immunological methods to detect the
S-gene products in the evening primrose (Oenothera
organensis), which has a sporophytic self-incompatibility
system. He raised antisera to pollen extracts of different
genotypes, and showed a precise correlation between S-genotype and particular antigens.

The success of this approach has encouraged other groups who have consistently been able to show a correlation between S-genotype and a specific antigen. For example, Nasrallah, Wallace and coworkers[32] raised antisera to homogenates of pistils of Brassica oleracea; the S-genotype specific antigen was partially characterized as a high isoelectric point glycoprotein. The antigen was extremely soluble and appeared at the same time as incompatibility developed, about three days prior to anthesis. In a more detailed study Nishio and Hinata[33] have shown that the antigen corresponding to genotype S_7 of Brassica campestris is a glycoprotein, isoelectric point pH 5.7, MW 57,000, with a carbohydrate to protein ratio of 1.2:1. The principal amino acids were serine, glutamic acid and glycine.

We have examined the style mucilage of Prunus avium in an attempt to define the components involved in expression of specificity and their relationship to the arabinogalactans detected in style mucilages of this and other plants.

In Prunus avium, all self-pollinations are incompatible, as are cross pollinations between varieties with the same S-genotype. In both incompatible and compatible pollinations, the initial events are apparently identical. The pollen tubes germinate and grow into the style. The cells of the transmitting tract of the style are joined by plasmodesmata into vertical files.[75] At maturity, the extracellular gel matrix becomes progressively less viscous, so that pollen tubes grow through an amorphous fluid matrix between quite separated files of cells. At some point within the style, growth of incompatible tubes is arrested. The precise zone of the style at which incompatible tube growth ceases, varies with ambient temperature (Lewis, 1942). Incompatible tubes swell at the tip and often burst; usually a deposit of aniline-blue staining material ("callose") is observed just behind the tip of the arrested tube. The behavior of pollen tubes during the growth, suggests that the reaction might involve contact between the growing pollen tube and the style secretions. We examined style secretions from two S-allele groups, S_1S_2 and S_3S_4 (Table 4).

Immunological approaches. By raising antisera to a pistil extract of styles of the S_3S_4 genotype (Lambert) we were able to demonstrate antigenic components which correlate

precisely with the S-allele group. In the homologous
reaction between antiserum and the diffusable material of
mature Lambert pistils, two distinct antigenic components
are separated* (Fig. 11). On immunodiffusion, there is an
outer band which correlates with the S-genotype (S-antigen)
and an inner band which is apparently specific to pistil
tissues (P-antigen) of members of the Rosaceae. On immunoelectrophoresis at pH 8.9 the S-antigen moves toward the
cathode and the P-antigen moves toward the anode.

When antiserum is tested with pistil diffusates from
other Prunus avium varieties of the same S-allele group
(Napoleon, Bing), an identical pattern is produced. However,
when the antiserum is tested with pistil diffusates from
varieties of a different S-genotype group (S_1S_2) the P-antigen is present but a band corresponding to the S-antigen is
barely detectable. When the antiserum is absorbed with the
eliciting antigen preparation or with the corresponding style
diffusates from other varieties of the same S-genotype group
(S_3S_4), and then tested, both S- and P-antigen are removed.
However, if the antiserum is absorbed with style diffusates
from the S_1S_2 group and tested, the P-antigen is completely
removed but the S-antigen remains (Table 4).

---

*These antigens, which correlate with the S-allele, can also
be detected using antisera raised to material secreted by
callus cells of Prunus avium, grown in liquid suspension
culture. Callus cells secrete a range of antigens into
their culture media: not only antigens specific to the
parental organ, but also antigens typical of other organs of
the parent plant.[66] In this study callus was raised from
leaf tissue of cv Napoleon (S_3S_4); the culture filtrate was
taken to 80% saturation with ammonium sulphate and antisera
was raised in rabbits to this protein fraction. A number
of antigens were detected, including the major antigen which
is specific to leaf tissues of the genus Prunus, and an
antigen apparently identical with the antigen found in pistil
extracts which correlates with the S-allele group. This
extraordinary finding that an antigen as specific in function
as the S-allele antigen of the pistil is secreted by callus
cells derived from tissues other than the pistil, dramatically illustrates the potential of callus cells for wide
expression of the plant genome.[67]

The S-antigen is also developmentally regulated. It is absent in early stages of style development - at mid balloon and petal show, but is present at late balloon and in mature pistils (24 hrs after flower opening). Also, it is present in both the upper and lower portions of the style but is absent from the ovary and receptacle. Thus, it is present at the right time (flower maturity), in the right place (style) to function in control of sterility. However, although there is a precise correlation of this antigen with S-genotype group, a direct function for this material has not been demonstrated.

Another interesting feature of this S-antigen is its precise tissue specificity. It could not be detected in pollen diffusates, ovary or receptacle of mature flowers or in leaf tissue of the S_3S_4 genotype. Nor could it be detected in a number of diffusates prepared from pistils of other members of the Rosaceae such as peach (Prunus persica) or rose (Rosa domestica).

Chemical approaches. These components have been isolated and partially characterized by traditional methods.

An extract of style collected from mature flowers of genotype (S_3S_4) was fractionated by ammonium sulphate precipitation. The individual fractions were examined for their capacity to interact with antiserum raised to the whole pistil extract.

Both the S and P antigens were concentrated in the 30 to 45% ammonium sulphate fraction and have been isolated on the basis of their differing charge properties, by DEAE-cellulose chromatography. They are quantitatively minor components, each accounting for less than 5% of the soluble material of the style (yield < 2 mg from 6,000 pistils). Both antigens are glycoproteins (S antigen 16.3%, P antigen 17.2% carbohydrate as glucose). There is no apparent immunological cross-reactivity, in that [^{125}I]-P antigen gave a single peak corresponding to molecular weight 32,000 when precipitated by its homologous antiserum, but no precipitation in the heterologous antiserum. [^{125}I]-S antigen gave two closely related peaks, apparent molecular weights, 37,000 and 39,000, when precipitated with its homologous

Table 4.   Antigens detected in style extracts of <u>Prunus</u> <u>avium</u> by immunodiffusion and immunoelectrophoresis.   Cultivars in S-allele group S_1S_2 were Bedford and Early Rivers; in group S_3S_4 were Bing, Lambert and Napoleon.

S-allele group of Pistil extract	S_3S_4		S_1S_2
Antiserum to extract of Lambert pistils (S_3S_4)	P	S	P (S faint)
Antiserum to extract of Lambert pistils (S_3S_4) absorbed with pistils of S-allele group (S_3S_4)	-	-	- -
Antiserum to extract of Lambert pistils absorbed with pistils of S-allele group (S_1S_2)	-	S	- -

antiserum, but no precipitation in the heterologous interactions (Mau, S-L, Raff, J and Clarke, A. E., unpublished observations).

The most remarkable feature of the analysis was the presence of a single major component: on the basis of the experience with <u>Gladiolus</u> and <u>Lilium</u> we expected to find an arabinogalactan, but found a high molecular weight (>70,000) macromolecule containing 95% protein, 5.4% carbohydrate (as glucose) with monosaccharides Glc:Man:Gal:Xyl in the ratio 5.3:1.4:2.2:1.0. The component was present in the 20 to 30% ammonium sulphate fraction. It was purified by G-200 Sephadex and Biogel P-6 chromatography and accounts for about 30% of the total soluble material of the whole style extract. As expected, from the monosaccharide analysis, there was no interaction with either the anti-arabinogalactan protein serum, or the β-glucosyl artificial carbohydrate antigen. However, whole homogenates of <u>Prunus</u> <u>avium</u> styles do interact with these probes, indicating the presence of arabinogalactans in the tissues. We located these arabinogalactans in the supernatant of the 85% ammonium sulphate fraction by their binding to both probes. In contrast to the arabinogalactans of <u>Gladiolus</u> and <u>Lilium</u>, they

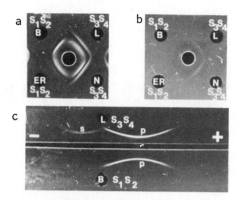

Figure 11.   S-Allele associations of antigens from the
pistil of the sweet cherry, Prunus avium.   (a) Immunodiffusion: antiserum in the inner well was raised to a style
extract from Lambert (S_3S_4).   The outer wells contained
antigens prepared from mature pistils of different varieties,
with known S-allele types: Lambert (L) and Napoleon (N)
belong to the S-allele group S_3S_4 and produce two bands with
the antisera.   Bedford (B) and Early Rivers (ER) of the
S-allele group S_1S_2 produce a single band continuous with
the inner band of the two produced with the S_3S_4 groups.

(b)   Immunodiffusion: The same antiserum (to S_3S_4)
was adsorbed with pistil extract from Bedford (S_1S_2 inner
well).   The outer wells contained antigens prepared from
mature pistils of different varieties.   The inner band,
common to both S_3S_4 and S_1S_2 extracts was removed by the
absorption.   The outer band was formed between the S_3S_4
varieties and the absorbed antisera.

(c)   Immunoelectrophoresis: The antiserum in the
centre trough was raised to a style extract from Lambert
(S_3S_4).   The upper wells contained antigen preparation from
Lambert (S_3S_4); two bands were produced.   The P band, was
common to pistils of the other S-allele group S_1S_2 and the
S band was only present in extracts of the S_3S_4 group.

were minor components of the Prunus style, accounting for
less than 2% of the total soluble material.   Alerted by
Larkins' observations[61] that protoplasts derived from many
cell types are agglutinated by β-glucosyl artificial antigen,
we also examined a membrane preparation from the style.   The
style homogenate was exhaustively extracted and washed with

buffer, and then extracted with 0.1% Nonidet-P40. This
detergent extract gave a precipitin band with both the anti-
arabionogalactan-protein serum and the β-glucosyl artificial
antigen.

In summary, the style extract of Prunus avium contains
a major single high molecular weight of glycoprotein component. There are two minor components which can be detected
by their antigenic properties; one is characteristic of the
S-genotype and is restricted to mature P. avium styles, the
other is present in the styles of all Prunus species tested.
The arabinogalactans, which are major components in
Gladiolus and Lilium styles, are minor components of the
soluble material in P. avium styles and are also present
in a membrane preparation.

An analysis of the aniline blue-staining material
produced as a result of self pollination of Secale cereale -
"callose" contains both (1→3)- and (1→4)-glucosidic linkages.

The third aspect of the pollen-stigma interaction which
we will consider is the deposition of aniline blue-staining
material as a response to incompatible matings in both the
gametophytic and sporophytic systems. For the sporophytic
system, these deposits are formed in the initiating pollen
tube and in stigmatic cells at the stigma surface. In the
gametophytic system they are formed in the tip of the pollen
tube within the style. This characteristic deposition is
such a reliable test that it is used horticulturally to establish relationships between potential breeding groups.
Material which gives the characteristic brilliant yellow
fluorescence with decolorized aniline blue, is generally
referred to as callose although the precise nature of the
material stained is not known. Because in one situation
(sieve tube callose of grape vine), the staining material
was shown to be a (1→3) β-glucan, it has been assumed that
a polysaccharide with the same structure is always involved
in this staining interaction. We have examined the nature
of the aniline blue staining material produced as a response
to incompatible self-mating in the grass Secale cereale.[68]

In this system, compatible pollen tubes penetrate the
stigma surface, grow intercellularly along the multicellular
papillae of the stigma, and enter the main stigma branch,
where they continue to grow intercellularly towards the

base of the stigma and into the ovary (Fig. 12). In incompatible matings, the pollen tubes grow intercellularly for
a distance of four or five cells in the stigmatic papillae
and growth is arrested after 45 minutes. Aniline blue-
staining droplets are detected 20 minutes after pollination
and by 1 hour have coalesced into distinct plugs. Deposition continues until 3 hours after pollination when the
material occludes about one third of the pollen grain surface and the basal part of the pollen tube, after which
there is no further deposition.

Under the conditions used for pollination, no wound
callose was produced. This was shown by the poor staining
of compatible pollen and tubes with aniline blue. Also,
when excised stigmas were treated with polystyrene beads (40
μm diameter) or sunflower pollen under the same conditions,
no wound callose was detected with aniline blue-staining.

To examine the nature of this incompatibility callose,
2,000 stigmas were self pollinated. Three hours later the
stigmas were homogenized; the callose deposits were released
as fragile solid plugs which fragmented on further homogeni-
zation. The deposits could not be separated by density gradient method and could only be solubilized under degradative
conditions. The approach of removing other components,
leaving te callose as a residue was adopted. Initially,
the homogenate was extracted with chloroform and methanol
to remove lipids from both the pollen tube and the outer
layer of the pollen grains, the pollenkitt. This was followed by a hot-guffer extraction to remove soluble cell wall
components and an acid chlorite treatment to oxidize sporo-
pollenin and any lignin. The total yield from 2,000 pollinated stigmas (dry weight 600 mg) was 80 mg. This material was further fractionated on the basis of its solubility in dimethyl sulphoxide (DMSO). The whole "callose"
preparation and both the DMSO soluble an insoluble fractions
retained the capacity to stain with aniline blue, characteristic of the deposits in the whole tissue.

Monosaccharide analyses of the "callose preparation
and both the DMSO soluble and insoluble fractions are shown
in Table 5. The whole preparation contained glucose as the
major monosaccharide as well as arabinose, xylose, galactose
and mannose. The DMSO insoluble fraction was extremely difficult to hydrolyze - only 25% of the total materail was

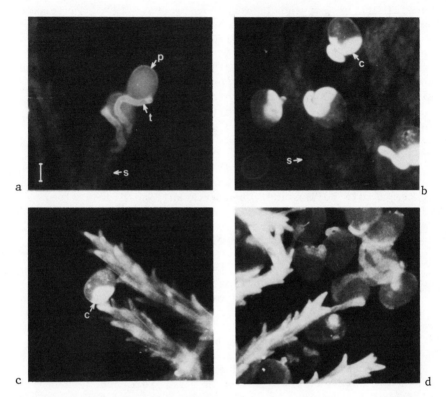

Figure 12.   Deposition of "callose", as a response to self
pollination in Secale cereale.
    Excised mature stigmas were pollinated by dipping in
fresh compatible or incompatible pollen. After 60 minutes
in a moist chamber at room temperature, the samples were
squashed, stained and examined by fluorescence microscopy
(magnification x 900; internal marker 25 μm).
    (a) Compatible pollen (p) showing the tube growth (t)
through the multicellular stigmatic hair (s); stained with
decolorized aniline blue. (b) Incompatible pollen on self
stigma (s) showing the comma-shaped callose deposit (c)
which occludes part of the pollen grain and the short pollen
tube; stained with decolorized aniline blue. (c) and (d)
as for (a) and (b) except the preparations were stained with
Calcofluor M2R New and viewed by fluorescence microscopy.
Stigma, pollen grains and tubes were all stained, however
the area corresponding to depositon of "callose" (c) is
intensely stained (Panel c).

recovered as reducing sugars and in the material, glucose was the major component with arabinose as a mino component.

Methylation analysis (Table 6) of this DMSO soluble fraction showed most of the glucose to be (1→4)-linked. However, as well a the expected terminal glycosyl residues, 9% of the material was recovered as a peak tentatively identified as 1,3,5 tri-O-acetyl-2,4,6-tri-O-methyl-$\underline{D}$-glucitol corresponding to 1,3-linked glycosyl residues. All the arabinose, which is present as a minor component, was recovered as terminal residues. Whether this material represents a mixture of a (1→4)-glucan and a (1→3)-glucan or a macromolecule containing both linkage or a mixture of all three types of polymer cannot be stated from the data. X-ray diffraction of the isolated "callose" gave a pattern similar to that of cellulose indicating that the predominant organizational feature in the "callose" preparation is linear runs of (1→4) β-glucosidic linkages. The absence of bands corresponding to the (1→3) β-glucan paramylon, also indicates the absence of long runs of (1→3)-linkages. Again these data can be interpreted as the presence of both cellulose and a glucan with runs of (1→3)-linkages, or of a single mixed linkage glucan.

This question was partially resolved by enzymic hydrolysis with purified, β-glucan hydrolases with defined linkage specificity. The results and the specificity of the enzymes are given in Table 7. The exo-acting (1→3)-glucan hydrolase released glucose from the whole callose preparation as well as both the DMSO-soluble and insoluble fractions - indicating that there are at least some chains with terminal (1→3)-β-glucosyl residues. The Rhizopus arrhizus endo-(1→3)-β-glucan hydrolase will cleave either (1→3)- or (1→4)-glucosidic linkages, joining 3-substituted glycosyl residues in a linear chain. Thus linkages in the sequences

$$-G3G\overset{|}{3}G- \text{ and } -G3G\overset{|}{4}G-$$
$$\downarrow \qquad\qquad \downarrow$$

will be hydrolyzed as shown. The products of hydrolysis of the whole "callose" preparation as well as the DMSO-soluble and insoluble fractions included laminaribiose as well as the mixed-linkage trisaccharide G4G3G. A similar pattern of products was produced by enzymic hydrolysis with the

Table 5. Monosaccharide analyses of the "callose preparation" from rye pollen after self-pollination, and the fractions obtained from the "callose preparation" by DMSO fractionation. Results are expressed as % (by weight) of total
carbohydrate.

| Monosaccharide | "Callose preparation" | DMSO fractionation | |
		Insoluble fraction	Soluble fraction
Glucose	61	26	90
Arabinose	18	58	10
Xylose	11	5	0
Galactose	6	7	0
Mannose	4	4	0

Table 6. Methylation analysis of the "callose preparation"
from rye pollen after self-pollination, and the fractions
obtained from this preparation by DMSO fractionation.

| Linkage type | Linkage composition (% peak area) | | |
| | "callose pre-paration" | DMSO fractionation | |
		Insoluble fraction	Soluble fraction
Terminal Ara$f$	7	7	5
1,3 linked Ara$f$	1	trace	0
Terminal Glc 1,5 linked Ara$f$	0	0	8
1,5 linked Ara$f$	11	11	0
Terminal Gal	6	7	0
1,3 linked Glc 1,2 linked Glc 1,2 linked Man	8	5	9
Xyl (? linked)	17	16	0
1,4 linked Glc	50	54	77

Bacillus subtilis (1→3):(1→4) β-glucan endo-hydrolase. This
enzyme will hydrolyze glucans containing both (1→3)- and
(1→4) β-glucans linkages in a linear sequence but will not
hydrolyze (1→3)-glucans or (1→4)-glucans which are homogeneous with respect to linkage type. This enzyme degraded
the DMSO-soluble fraction most effectively, but also released trace amounts of oligosaccharides from the DMSO-
insoluble and whole "callose" preparation. The major
products were G4G3G and G4G4G3G.

The Streptomyces enzyme gave the same products with
all three callose fractions: cellobiose (G4G), cellotriose
(G4G4G), cellotetraose (G4G4G4G), and the mixed-linkage
trisaccharide (G3G4G). The specificity of this enzyme is
for (1→4)-glucosidic linkages where the glycosyl residue is
4-substituted. Thus, it will hydrolyze either (1→4)
β-glucans such as cellulose or the linear heteropolymers
such as barley glucan. The findings, that each of the three
endoacting enzymes produced mixed linkage trisaccharides
from the whole callose preparation and from the DMSO-soluble
and insoluble fractions indicates that the (1→3)- and (1→4)-
linkages must be present in the same molecule.

Thus the "callose" produced in response to self pollination of Secale cereale contains linear glucose polymers
which have (1→3)-β and (1→4)-β-linkages in the same molecule; the evidence available does not rule out the possibility that this preparation may also contain substantial
amounts of the (1→4)-β-glucan cellulose, and/or some homogeneous (1→3)-β-glucan, and poses the question, 'Are all
macromolecules which give positive fluorescence with
aniline blue (1→3)-β-glucans?'.

The question is difficult to resolve, because the
material in the cell or tissue may react differently when it
is isolated - that is, the secondary and tertiary structure
and the hydration state as well as the primary sequence of
monosaccharide, may influence the interactions. The available evidence indicates that whereas glucans containing high
proportions of (1→3)-linkages are usually stained, glucans
with other linkage types may also stain. Furthermore, other
cell wall components such as cellulose and mixed (1→3):(1→4)-
linkage glucans also bind the fluorochrome.[69,70,71] The
fluorescent brightener, Calcofluor also stained the callose
deposits in Secale stigmas more intensely than the other cell

Table 7. Tentative identification of the products from enzyme hydrolysis of "callose preparation" and the DMSO insoluble and DMSO soluble fractions.

Enzyme treatment	Linkage specificity	Products of enzyme treatment identified by paper chromatography		
		Callose* preparation	DMSO insoluble* fraction	DMSO soluble fraction
Euglena gracilis (EC 3.2.1.58) (exo 1,3 ase)	G3G→	G (trace)	G (trace)	G (oligomers also)
Rhizopus arrhizus (EC 3.2.1.6) (endo 1,3 ase)	- G3G3G - or - G3G4G -	G (trace) G3G G4G3G	G (trace) G3G G4G3G	G (strong) G3G G4G G4G3G
Bacillus subtilis (EC 3.2.1.73) (1,3;1,4 ase)	- G4G3G4G -	G (trace) G4G3G (trace)	G (trace) G4G3G (trace)	G G3G G4G G4G3G G4G4G3G
Streptomyces (EC 3.2.1.4) (1,4 ase)	- 4G4G -	G4G G3G4G ? G4G4G G4G4G4G	G4G G3G4G ? G4G4G G4G4G4G	G4G G3G4G ? G4G4G G4G4G4G

*After incubation with the enzyme, there was an insoluble residue.

wall regions: this dye binds strongly to cellulose,[72,73]
but will also bind mixed linkage (1→3):(1→4)-glucans.[73,74]
Certainly these dyes, aniline blue and Calcofluor, as well
as the β-glucosyl artificial carbohydrate antigen are useful
cytochemical probes: for each dye we can define the major
class of compounds stained, but not the precise physico-
chemical requirements for staining.

Aniline blue-staining material is also found in pollen
tube walls.   Herth and coworkers[76] examined the nature of
Lilium pollen tube wall material and found both (1→3) and
(1→4) β-glucosidic linkages in an isolated wall fraction.
They faced a similar dilemma as to whether these originated
from a mixture of (1→3)- and (1→4)-β-glucans, or a mixed
linkage (1→3):(1→4)-β-glucan.   The staining of pollen tube
walls with resorcin blue, another "callose specific" stain,
is abolished by treatment with 1,3 β-glucan hydrolase indicating the presence of (1→3)-glucosidic linkages.[77]

Observations on "callose" formed in response to an incompatible fungal infection support the idea that aniline
blue-staining indicates, but is not necessarily restricted
to (1→3)-β-glucans.   Collars of aniline blue staining material "callose", are laid down on walls of Zea mays root
epidermal cells as a response to contact with invading
hyphae of Phytophthora cinnamomi.   In sections of infected
root, fluorescence with aniline blue is abolished by digestion of the section with (1→3)-β-glucan exo-hydrolase (from
Euglena gracilis) (1→3)-β-glucan endo-hydrolase (from
Rhizopus arrhizus) and (1→3):(1→4)-β-glucan endo-hydrolase
(from Bacillus subtilis) but not (1→4)-β-glucan endo-hydrolase (from Streptomyces sp.).   Also the "callose" collars
give a positive stain with periodate-Schiff reagent indicating the presence of linkages other than (1→3)-β-gluco-
sidic linkages.   However, although the capacity to stain
with aniline blue is abolished by enzymic digestion, the
"callose" collars retain their capacity to stain with the
periodate-Schiff reagent.[78]

Other similarities can be seen between pollen tube
growth through the style canal and fungal infection of roots
(Fig. 13) (Table 8).   In each case the host tissues are
penetrated by a genetically foreign invader (pollen tube or
fungal hypha).   In both cases, the invader may be accepted
and grow through the tissue, or the growth may be arrested.

In instances where there are highly specific interactions
such as self-incompatible pollinations or race-specific
pathogenesis, the control is through either a single gene
or a few gene loci. This recognition seems to depend on an
interaction between surface secretions and their complementary receptors. In both cases, there are a number of responses to the interaction, among which is the deposition
of "callose". The precise nature of the molecules which
elicit this and other responses and of the receptors for
these signals, is just now beginning to be unravelled.

CONCLUSIONS

1. The study of the events of pollination at the molecular level is a new field of cell biology. The groundwork
has been laid by the careful microscopic observations of
pollination biologists. These studies have pointed to the
stigma surface and style secretions as likely candidates
for mediating recognition between the pollen-grain or tube
and the female sexual tissues.

2. The nature of the components present at the interacting
surfaces can be partially established by classic analytical
techniques. Immunological techniques have unexploited
potential for analysis of minor components of plant cell
surfaces. The use of these techniques has allowed an in-
sight into the nature and organization of animal cell surface components, and there is no doubt that immunologicial
methods will be of great value in future studies of plant
cell surfaces.

3. A major obstacle in studies of cell recognition during
pollination is the development of a rapid quantitative assay
for compatible and incompatible pollination. In vivo assays
which depend on seed set are time-consuming and may be
subject to environmental hazards. Assays for pollen tube
growth usually depend on subjective observations of the
numbers of tubes in a style canal. Assays which depend on
the interaction between pollen tubes grown in vitro and
particular components have some potential, although there
are ultrastructural differences between pollen tubes growth
in vivo and in vitro.

4. It seems likely that the recognition events of pollination will be based on the same mechanisms operative for

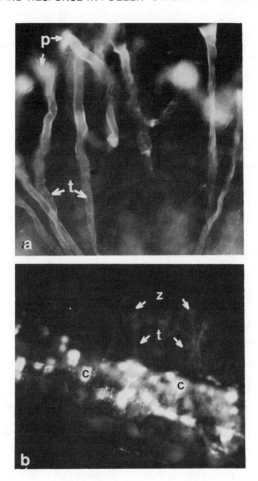

Figure 13.   Fluorescent micorgraphs showing a comparison of
pollination and fungal infection.   (a)   The pollen grains
(p) at the surface of the stigma produce tubes (t) which
grow through the style canal.   (b)   The zoospores (z) produce germ tubes (t) and hyphae which grow through the root
tissue.   The host responds by deposition of callose (c).

recognition between higher plant cells in other situations.
That is, receipt of extracellular information, either at
the cell wall surface or at the plasma membrane after diffusion through the wall.   One response to recognition of

Table 8. Comparison of self-incompatible pollination and
race specific pathogenesis in terms of the possible recognition events and response. The ultimate response is arrest
of growth of the pollen tube or fungal hypha. One of many
facets of the total response is deposition of callose in
the pollen tube or cell walls of the infected tissue.

	SIGNAL	RECEPTOR	RESPONSE
Gametophytic self-incompatible pollination	soluble S-gene product in style	pollen-tube (wall? plasma membrane?)	Arrest of pollen tube growth ↓ ↓ callose in pollen tube
Race specific pathogenesis	secretion of fungal hyphae	host plant (plasma membrane?)	Arrest of fungal growth ↓ ↓ callose in host wall

incompatible pollen is production of callose, identified by
its fluorescent staining with aniline blue. The material
is also produced as a response to many fungal infections:
its formation is also part of the general wound response.

5.   Analytical work has so far only been undertaken for a
few components of a few systems, there is insufficient data
available to draw any generalizations regarding the molecular basis of the recognition events.

6.   There is no information available regarding the putative
receptors at the pollen tube surface. The nature of the
secreted components of the stigma and the pistil have been
examined for the monocotyledons Gladiolus and Lilium. Both
contain a variety of components, among which is an arabino-
$(1{\to}3):(1{\to}6)$-galactan which accounts for about half the total
non-dialyzable material of the secretion. Related

arbinogalactans are present in style extracts of other
species. The role these components or indeed the other
minor components play in pollination is not known.

7. For a number of systems, components which correlate with
S-genotype have been demonstrated; these components have an
implied, but not proven, role in recognition of compatible
pollen.

ACKNOWLEDGEMENTS

We would like to acknowledge the invaluable contributions made by Dr. V. Vithanage, Mr. John Raff and Ms. S-L.
Mau to the experimental work described in this paper as well
as for their thoughtful discussions during the progress of
the work. We are also indebted to Prof. R. B. Knox for his
support and advice, particularly with the biological aspects
of the work. We would also like to thank Prof. B. A. Stone
for his generous gifts of enzymes, for discussion and for
critical review of the manuscript. Finally, we are most
grateful to Dr. Elizabeth G. Williams, Grasslands Division,
DSIR, New Zealand for her constructive advice on the genetic
aspects of the work, and her careful reading of the manuscript. This work was supported by a grant from the
Australian Research Grants Commission.

REFERENCES

1. Clarke, A. E. and R. B. Knox. 1978. Cell recognition
    in flowering plants. Q. Rev. Biol. 53:3-28.
2. Frazier, W. and L. Glaser. 1979. Surface components
    and cell recognition. Annu. Rev. Biochem. 48:491-523.
3. Sharon, N. 1979. Possible functions of lectins in
    micro-organisms, plants and animals. In Glyconjugate
    Research. 1. (J. D. Gregory and R. W. Jeanloz, eds.).
    Academic Press, New York. pp. 458-491.
4. Williams, A. F. 1978. Membrane glycoproteins in
    recognition. Biochem. Soc. Trans. 6:490-494.
5. Barondes, S. H. and S. D. Rosen. 1976. Cell surface
    carbohydrate-binding proteins: Role in cell recogni-
    tion. In Neuronal Recognition. (S. Barondes, ed.).
    Plenum Press, New York. pp. 331-356.
6. Goldstein, I. J., R. C. Hughes, M. Monsigny, T. Osawa
    and N. Sharon. 1980. What should be called a lectin.
    Nature 285:66.

7. Greaves, M. F. 1975. Cellular Recognition. Outline
    Studies in Biology. Chapman and Hall, London.
8. Goldstein, I. J. and C. E. Hayes. 1978. The lectins:
    carbohydrate-binding protein of plants and animals.
    Adv. Carbohydr. Chem. Biochem. 35:128-340.
9. Burke, D., L. Mendonca-Previato and C. E. Ballou. 1980.
    Cell-cell recognition in yeast. Purification of
    Hansenula wingei 21-cell sexual agglutination factor
    and comparison of the factors from three genera. Proc.
    Natl. Acad. Sci. USA 77:318-322.
10. Hughes, R. C. 1979. Cell surface carbohydrates in
    relation to receptor activity. In Glycoconjugate
    Research. Vol. 2. (J. D. Gregory and R. W. Jeanloz,
    eds.). Academic Press, New York. pp. 986-1003.
11. Weise, L. 1974. Nature of sex-specific glycoprotein
    agglutinins in Chlamydomonas. Ann. New York Acad. Sci.
    234:283-295.
12. Bolwell, G. P., J. A. Callow, M. E. Callow and L. V.
    Evans. 1979. Fertilization in brown algae. II.
    Evidence for lectin-sensitive complementary receptors
    involved in gamete recognition in Fucus serratus. J.
    Cell Sci. 36:19-30.
13. Morris, E. R., D. A. Rees, G. Young, M. D. Walinshaw
    and A. Darke. 1977. Order-disorder transition for a
    bacterial polysaccharide in solution. A role for
    polysaccharide conformation in recognition between
    Xanthomonas pathogen and its plant host. J. Mol. Biol.
    110:1-16.
14. McNeil, M., A. G. Darvill and Albersheim. 1979. The
    structural polymers of the primary cell walls of dicots.
    Prog. Chem. Org. Nat. Prod. 37:191-249.
15. Carpita, N., D. Subularse, D. Monteginos and D. P.
    Delmer. 1979. Determination of the pore size of cell
    walls of living plant cells. Science 205:1144-1147.
16. Albersheim, P. and B. S. Valent. 1978. Host-pathogen
    interaction in plants. J. Cell Biol. 78:627-643.
17. Peters, B. M., D. H. Cribbs and D. A. Stelzig. 1978.
    Agglutination of plant protoplasts by fungal cell wall
    glucans. Science 201:364-365.
18. Clarke, A. E. and R. B. Knox. 1980. Plants and
    Immunity. Dev. Comp. Immunol. 3:571-589.
19. Albersheim, P. and A. J. Anderson-Prouty. 1975.
    Carbohydrates, proteins, cell surfaces and biochemistry
    of pathogens. Annu. Rev. Plant Physiol. 26:31-52.

20. Liener, I. E. 1976. Phytohemagglutinins (Phytolectins).
     Annu. Rev. Plant Physiol. 27:291-319.
21. Callow, J. A. 1977. Recognition, resistance and the
     role of plant lectins in host-parasite interactions.
     Adv. Bot. Res. 4:1-49.
22. Sequeira, L. 1978. Lectins and their role in host
     pathogen specificity. Annu. Rev. Phytopathol.
     16:435-81.
23. Schmidt, E. L. 1979. Inititation of plant root-
     microbe interactions. Annu. Rev. Microbiol.
     33:355-376.
24. Gunning, B. E. S. and A. W. Robards. 1976.
     Plasmodesmata: Current knowledge and outstanding
     problems. In Intercellular Communication in Plants:
     Studies on plasmodesmata. (B. E. S. Gunning and A. W.
     Robards, eds.). Springer-Verlag, New York. pp. 297-311.
25. Dickinson, H. G. and D. Lewis. 1973. Cytochemical and
     ultrastructural differences between intraspecific
     compatible and incompatible pollinations in Raphanus.
     Proc. Roy. Soc. (Lond.) Ser.B. 183:21-38.
26. Knox, R. B. 1973. Pollen-wall proteins: pollen-
     stigma interactions in ragweed and Cosmos (Compositae)
     J. Cell Sci. 12:421-443.
27. Heslop-Harrison, J., R. B. Knox, Y. Heslop-Harrison and
     O. Mattsson. 1975. Pollen-wall proteins: emission
     and role in incompatibility responses. In The biology
     of the male gamete. (J. G. Duckett and P. A. Racey,
     eds.). pp. 189-202. [Suppl. 1 to the Biol. J. Linnean
     Soc. Vol. 7].
28. Knox, R. B. and A. E. Clarke. 1980. Discrimination
     of self and non-self in plants. Contemp. Top.
     Immunobiol. 9:1-30.
29. Nettancourt, D. 1977. Incompatibility in Angiosperms.
     Springer-Verlag, Berlin and New York.
30. Dickinson, H. G. and D. Lewis. 1973. The formation
     of the tyrphine coating the pollen grains of Raphanus,
     and its properties relating to the self-incompatibility
     system. Proc. Roy. Soc. Lond. Ser.B. 184:149-65.
31. Heslop-Harrison, J., R. B. Knox and Y. Heslop-Harrison.
     1974. Pollen-wall proteins: exine-held fractions
     associated with the incompatibility response in the
     Cruciferae. Theor. Appl. Genetics. 44:133-137.
32. Nasrallah, M. E. and D. H. Wallace. 1967. Immuno-
     gentics of self-incompatibility in Brassica oleracea.
     Heredity 22:519-527.

33. Hinata, K. and T. Nishio. 1978. S-Allele specificity
     of stigma proteins in Brassica oleracea and B.
     campestris. Heredity 41(1):93-100.
34. Nettancourt, D. de, M. Devreux, A. Buzzini, M. Cresti,
     E. Pacini and Sarfatti. 1973. Ultrastructural aspects
     of the self-incompatibility mechanisms in Lycopersicum
     peruvianum. J. Cell Sci. 12:403-419.
35. Raff, J. 1980. Unpublished observations.
36. Heslop-Harrison, J. 1978. Genetics and physiology of
     angiosperm incompatibility systems. Proc. Roy. Soc.
     Lond. Ser.B. 202:73-92.
37. Knox, R. B. 1979. Pollen and Allergy. Institute of
     Biology Studies in Biology No. 107. Edward Arnold,
     London.
38. Mascarenhas, J. P. 1975. The biochemistry of Angio-
     sperm pollen development. Bot. Rev. 41:259-314.
39. Loewus, F. and C. Labarca. 1973. Pistil secretion
     product and pollen tube wall formation. In Biogenesis
     of Plant Cell Wall Polysaccharides. (F. Loewus, ed.).
     Academic Press, New York. pp. 175-193.
40. Knox, R. B. 1981. Pollen-Stigma Interactions. In
     Encyclopaedia of Plant Physiology (New Series): Plant
     carbohydrates. (F. A. Loewus and W. Tanner, eds.).
     Springer-Verlag, Berlin and New York.
41. Lewis, D. 1952. Serological reactions of pollen
     incompatibility substances. Proc. Roy. Soc. Lond.
     Ser.B. 140:127-135.
42. Morre, D. J. and W. J. Van der Woude. 1974. Origin
     and growth of cell surface components. In Macromole-
     cules relating growth and development. (E. D. Hay,
     T. J. King, J. Papaconstatinou, eds.). 30th Symp. Soc.
     Dev. Biol. New York and London, Academic Press.
     pp. 81-111.
43. Knox, R. B., A. Clarke, S. Harrison, P. Smith and J. J.
     Marchalonis. 1976. Cell recognition in plants:
     Determinants of the stigma surface and their pollen
     interactions. Proc. Natl. Acad. Sci. USA 73:2788-2792.
44. Clarke, A., P. Gleeson, S. Harrison and R. B. Knox.
     1979. Pollen-stigma interactions: Identification and
     characterization of surface components with recognition
     potential. Proc. Natl. Acad. Sci. USA 76:3358-3362.
45. Clarke, A. E., A. Abbott, T. Mandel and J. Pettitt.
     1980. Organization of the wall layers of the stigmatic
     papilli of Gladiolus gandavensis: a freeze fracture
     study. J. Ultrastruct. Res. 73:(In Press).

46. Yariv, J., M. M. Rapport and L. Graf. 1962. The
     interaction of glycosides and saccharides with antibody
     to the corresponding phenylazo glycosides. Biochem. J.
     85:383-388.
47. Yariv, J. H. Lis and E. Katchalski. 1967. Precipita-
     tion of arabic acid and some seed polysaccharides by
     glycosylphenylazo dyes. Biochem. J. 105:1C-2C.
48. Jermyn, M. A. and Y. M. Yeow. 1975. A class of
     lectins present in the tissues of seed plants. Aust.
     J. Plant Physiol. 2:501-531.
49. Clarke, A. E., P. A. Gleeson, M. A. Jermyn and R. B.
     Knox. 1978. Characterization and localization of
     β-lectins in lower and higher plants. Aust. J. Plant
     Physiol. 5:707-722.
50. Jermyn, M. A. 1978. Comparative specificity of
     Concanavalin A and the β-lectins. Aust. J. Plant
     Physiol. 5:687-696.
51. Clarke, A. E., R. L. Anderson and B. A. Stone. 1979.
     Form and function of arabinogalactans and arabino-
     galactan-proteins. Phytochemistry 18:521-540.
52. Gleeson, P. A. and M. A. Jermyn. 1979. Alteration in
     the compositon of β-lectins caused by chemical and
     enzymic attack. Aust. J. Plant Physiol. 6:25-38.
53. Gleeson, P. A. and A. E. Clarke. 1979. Structural
     studies on the major component of the Gladioulus style
     mucilage, an arabinogalactan-protein. Biochem. J.
     181:607-621.
54. Luft, J. H. 1976. The structure and properties of the
     cell surface coat. Int. Rev. Cytol. 45:291-382.
55. Gleeson, P. A., M. A. Jermyn and A. E. Carke. 1979.
     Isolation of an arabinogalactan-protein by β-lectin
     affinity chromatography on tridacnin-Sepharose 4B.
     Anal. Biochem. 92:41-45.
56. Gleeson, P. A. and A. E. Carke. 1980. Comparison of
     the structures of the major components of the stigma
     and style secretions of Galdiolus: the arabino-3,6-
     galactans. Carbohydr. Res. 83:187-192.
57. Aspinall, G. O. and K. G. Rosell. 1978. Polysac-
     charide component in the stigmatic exudate from Lilium
     longiflorum. Phytochemistry. 17:919-921.
58. Kristen, U., M. Biedermann, G. Liebezeit and R. Dawson.
     1979. The composition of stigmatic exudate and the
     ultrastructure of the stigma papillae in Aptenia
     cordifolia. Europ. J. Cell Biol. 19:281-287.

59. Clarke, A. E. and B. A. Stone. 1982. Chemistry and
    biology of 1,3-β-D-Glucans. Macmillan, London.
60. Gleeson, P. A. and A. E. Clarke. 1980. Antigenic
    determinants of a plant proteoglycan, the Gladiolus
    style arabinogalactan-protein. Biochem. J.
    191:437-447.
61. Larkin, P. J. 1977. Plant protoplasts, agglutination
    and membrane bound β-lectins. J. Cell Sci. 26:31-46.
62. Gleeson, P. A. and A. E. Clarke. 1980. Arabino-
    galactans of sexual and somatic tissus of Gladiolus and
    Lilium. Phytochemistry 19:1777-1782.
63. Anderson, D. M. W. and I. C. M. Dea. 1969. Chemo-
    taxonomic aspects of the chemistry of Acacia gum
    exudates. Phytochemistry 8:167-176.
64. Labarca, C. and F. Loewus. 1972. The nutritional role
    of pistil exudate in pollen tube wall formation in
    Lilium longiflorum. I. Utilization of injected stig-
    matic exudate. Plant Physiol. 50:7-14.
65. Labarca, C. and F. Loewus. 1973. The nutritional role
    of pistil exudate in pollen tube wall formation in
    Lilium longiflorum. II. Production and utilization of
    exudate from stigma and stylar canal. Plant Physiol.
    52:87-92.
66. Raff, J. W., J. Hutchinson, R. B. Knox and A. E. Clarke.
    1979. Cell recognition: antigenic determinants of
    plant organs and their cultured callus cells.
    Differentiation 12:179-186.
67. Raff, J. W. and A. E. Clarke. 1980. Characterization
    of specific antigen secreted by suspension cultured
    callus cells of the sweet cherry Prunus avium. Planta
    (submitted).
68. Vithanage, H. I. M. V., P. A. Gleeson and A. E. Clarke.
    1980. Callose: its nature and involvement in self-
    incompatibility response in Secale cereale. Planta
    148:498-509.
69. Faulkner, G., W. C. Kimmins and R. G. Brown. 1973.
    The use of fluorochromes for the identification of
    β-1,3 glucans. Can. J. Bot. 51:1503-1504.
70. Fulcher, R. G., G. Setterfield, M. E. McCully and P. J.
    Wood. 1977. Observations on the aleurone layer. II.
    Fluorescence microscopy of the aleurone-sub-aleurone
    junction with emphasis on possible β-1,3-glucan depo-
    sition in barley. Aust. J. Plant Physiol. 4:917-928.

71. Smith, M. M. and M. E. McCully. 1978. A critical
     evaluation of the specificity of aniline blue induced
     fluorescence. Protoplasma 95:229-254.
72. Hughes, J. and M. E. McCully. 1975. The use of an
     optical brightener in the study of plant structure.
     Stain Technol. 50:319-329.
73. Wood, P. J. 1980. Specificity in the interaction of
     dyes with polysaccharides. Ind. Eng. Chem. Prod. Res.
     Dev. 19:19-23.
74. Wood, P. J. and R. G. Fulcher. 1978. The interaction
     of some dyes with cereal β-glucans. Cereal Chem.
     55:952-966.
75. Cresti, M., F. Ciampolini, E. Pacini, K. Ramulu and
     Devreux. 1978. Gamma irradiaton of Prunus avium
     flower buds: effects on stylar development - an ultra-
     structural study. Acta Bot. Neerl. 27:97-106.
76. Herth, W., W. W. Franke, H. Bittiger, R. Kuppel and
     G. Keilich. 1974. Alkali-resistant fibrils of β-1,3-
     and β-1,4-glucans: Structural polysaccharides in the
     pollen tube wall of Lilium longiflorum. Cytobiologie
     9:344-367.
77. Reynolds, J. D. and W. V. Dashek. 1976. Cytochemical
     analysis of callose localization in Lilium longiflorum
     pollen tube. Ann. Bot. 40:409-416.
78. Hinch, J. M. and A. E. Clarke. 1980. Callose forma-
     tion as a response to infection of Zea mays roots by
     Phytophthora cinnamomi. Physiol. Plant Pathol.
     (submitted).

Chapter <u>Nine</u>

BACTERIAL ATTACHMENT TO PLANT CELL WALLS

MARIAMNE H. WHATLEY AND LUIS SEQUEIRA

Department of Plant Pathology
University of Wisconsin-Madison
Madison, WI   53706

INTRODUCTION

     A plant encounters a large number of potential patho-
gens in its environment, but, because of the highly specific nature of most host-pathogen interactions, rarely
does successful infection occur. This specificity apparently is dependent on the initial recognition between
the plant and pathogen, which may be mediated by the interaction of complementary macromolecules on the surfaces of
both organisms. Recognition can facilitate growth of both
organisms, as is the case of symbiotic relationships.
Recognition also could function as a defense mechanism.
A plant can recognize and immobilize a potential pathogen,
thus preventing its multiplication. Though the hypothesis
of recognition as a specific defense mechanism is an attractive explanation of various resistance phenomena, it has
not been demonstrated unequivocally. Most of the work in
this area involves symbiotic or plant pathogenic bacteria.
This paper will examine the evidence for attachment of bacteria to plant cell walls and then proceed to a discussion
of the nature of the bacterial and plant components that
may be involved.

EVIDENCE FOR BACTERIAL ATTACHMENT TO PLANT CELL WALLS

When bacteria invade a plant, the plant often responds actively. For this response to occur, there must be
close proximity between the bacterium and host cell wall.[1]
Other than in the case of infection by soft rot bacteria,
in which plant material is enzymatically disintegrated
ahead of the pathogen, bacteria generally interact with
the host. In the cases of crown gall tumor induction,
in which plasmid transfer occurs between Agrobacterium and
its host, and root nodule formation, in which nitrogen-
fixing bacteria are encapsulated in the root hair, it is
clear that close association between host and parasite is
necessary. In these cases, the cell wall, though it is
sometimes thought of as an inert barrier, is very much involved in the host response. Incompatible and saprophytic
bacteria also stimulate a host response. Though the need
for specific recognition and attachment is apparent in the
case of infections by crown gall and nitrogen-fixing bacteria, there is still some controversy as to the role of
specific attachment in other interactions. The following
section reviews and discusses evidence for specific attachment in plant-bacterial interactions.

Agrobacterium tumefaciens

Infection of a susceptible host by Agrobacterium
tumefaciens results in non-self limiting tumors. A wound
is necessary for tumor induction. Once the induction
process is complete, the presence of the bacteria is no
longer required for the continued growth of the tumor.
By means of a pinto bean leaf tumor bioassay, Lippincott
and Lippincott[2] showed that bacterial attachment to a
plant host wound site is a necessary first step in tumor
induction. Tumor initiation by virulent strains of A.
tumefaciens is inhibited by the presence of either heat-
killed virulent Agrobacterium cells or certain living
avirulent cells. Some avirulent strains of Agrobacterium
show no inhibitory effect, however. Reduction in tumor
number occurs if the inhibitory cells are added with or 15
minutes before the virulent inoculum, but not if added 15
minutes later. It appears, therefore, that the binding
process is complete in 15 minutes.[2]

Binding of Agrobacterium has been shown to occur in
other hosts besides pinto beans.[3][6] Electron microscopy
has also provided evidence for Agrobacterium-plant attachment. Bogers[7] and Schilperoort[3] showed a close association between the virulent bacterium and the plant cell
wall, with apparent envelopment of the pathogen by material
that appeared to be of host origin. Agrobacterium also
binds to carrot cells in suspension culture[8] and to Datura
cells in suspension.[9]

The necessity of binding between Agrobacterium and
its host has become more significant since the discovery
that a portion of the Ti plasmid is actually transferred
to and transcribed in the host cell.[10,11] For this transfer to occur, close contact probably is necessary since
DNA would have to move through the bacterial envelope and
plant cell wall during the course of tumor induction. The
binding and transfer processes may involve several steps,
of which initial recognition may be only one. Many aspects
of cell binding, including changes in surfaces that would
allow plasmid transfer, remain to be investigated.

Besides causing tumor induction, A. tumefaciens has
also been shown to speed up normal plant development. In
the moss Pylaisiella selwynii, the presence of virulent,
viable A. tumefaciens speeds up gametophore induction and
increases the number of gametophores.[12] This process requires physical contact between the moss protonema and the
bacteria.[13] Scanning electron microscopy also has been used
to document the close moss-bacterial cell association.[14]
The germ tube appeared to be the site where effective attachment may occur. The different species of Agrobacterium
tested adhered with different orientations, A. tumefaciens
cells attaching lengthwise on the filament. Though a non-
site binding strain of A. radiobacter also adhered, it had
an end-on orientation and binding appeared to be mediated
by the flagellae. This difference in adherence may be
responsible for the ineffectiveness of this strain on gametophore development.

Rhizobium-legume interactions

As with the closely related Agrobacteria, it is clearly
important to have a close association between Rhizobia and
their hosts during the nodulation process. The initial

steps in this process involve deformation of the root hair, invagination of the root hair plasmalemma, and formation of an infection thread. There is a great deal of specificity in these processes, since most Rhizobia will nodulate only one host. Though specificity theoretically could be determined at many points in the long and complex process of nodulation, it appears that the initial recognition and binding may be the key determinant. It has been suggested that attachment is not the basis of host-symbiont specificity because some Rhizobia have been shown to attach strongly to non-hosts.[15] Although these data suggest that simple attachment is not necessarily sufficient to determine specificity, they do not disprove attachment as a prerequisite for nodulation.

When observed by electron microscopy, Rhizobia appeared to attach in and end-on, polar fashion; there was production of microfibrillar material, identified as cellulose, at the site of attachment.[16] Similar microfibrils are produced by the bacteria in culture and the authors suggested that continued production during the infection process may help mediate adsorption to the root hair surface. The possibility that the surface fibrillar material is of host origin is not entirely ruled out, however. The authors also concluded that the outer fibrillar layer of the infection thread is cellulose of host origin.

Dazzo et al.[17] compared attachment of non-nodulating Rhizobium and nodulating R. trifolii to clover root hairs quantitatively and found attachment was four to five times greater with the latter.

Pseudomonas spp.

When large populations of incompatible strains of Pseudomonas solanacearum are introduced into tobacco leaves, there is a rapid hypersensitive response (HR) which is characterized by collapse of plant cells and reduction in bacterial numbers.[18] Compatible bacteria, however, do not induce the HR, multiply rapidly, and spread from the area of infiltration.

Sequeira et al.[19] studied the ultrastructural changes associated with the introduction of virulent, avirulent or incompatible strains of P. solanacearum into tobacco leaves.

The avirulent or incompatible bacteria are attached to the
cell walls and enveloped by fibrillar and granular host
material. Also, there are changes in the host cell wall.
A pellicle, probably of cuticular origin, separates from
the and surrounds the bacteria. The plasmalemma separates
from the cell wall and there is an accumulation of membrane-
bound vesicles in the space between the plasmalemma and the
cell wall. The virulent cells, however, are not attached
and remain free to multiply in the intercellular fluid.
There is some breakdown of cell wall structure, apparently
caused by proteolytic and cellulolytic enzymes, but no
extensive changes in organelle structure for 48 hours or
longer. Heat-killed avirulent cells bind in the same
manner as live ones, but do not induce the cell collapse
associated with the HR. The HR can be prevented by pretreatment with dead cells; live cells infiltrated 24 hours
later will not attach and the HR is not induced.

Goodman et al.[20] obtained similar results with the inoculation of P. pisi and saprophytic species of Pseudomonas
into tobacco leaves. They reported the separation of the
wall cuticle from the cell wall at the point of attachment
of the incompatible bacteria. The cuticle became progressively thicker and contained fibrillar material and large
numbers of vesicles. The bacteria near the plant cell wall
appeared to be enveloped and immobilized by the cuticle.
Part of the enveloping material seemed to involve plant
contents which had migrated out from the cell wall. The
saprophytic bacteria also induced separation of the cuticle.
Compatible bacteria, P. tabaci, induced aggregation of some
of the loose fibrillar material, but none of the vesicles
that were apparently involved in the formation of the
thickened cuticle.

Sing and Schroth[21] reported initially that saprophytic
P. putida cells infiltrated into "Red Kidney" bean leaves
were enveloped by fibrillar material of apparently plant
origin after the initial attachment between bacteria and
host cell walls. The cuticular material was described as
less granular than in tobacco. Pathogenic strains, either
compatible or incompatible, did not attach and were not
enveloped. Unlike the tobacco system, attachment did not
appear necessary for induction of the HR since P. tomato
did not attach, but caused the HR.

A more recent paper from the same laboratory suggested that the apparent attachment and envelopment is an
artifact.[22] They suggested that when the intercellular
spaces of bean leaves are infiltrated, the water dissolves
materials from the cell surface. As the fluid recedes,
due to evaporation or imbibition, the dissolved material
forms a film at the water-air interfaces. Infiltrated
bacteria are physically trapped by this film. Attachment
of saprophytes was prevented by watersoaking the leaves
continuously. They also observed that films were associated
not necessarily with bacteria, but with the interstices between plant cells. In addition, they suggested the film
may aid pathogenesis by maintaining a liquid phase around
the bacterial cell.

Obviously, this paper raises several important criti-
cisms of the electron microscopic evidence for bacterial
attachment. If the attachment and envelopment observed in
tobacco result from physical drying rather than an active
host response, there must be a re-evaluation of the specificity of attachment and its role as a defense mechanism.

It should be remembered that the results obtained
with bean are not necessarily applicable to those with tobacco or other hosts. The nature of the envelopment is
visibily different in the two hosts.[21, 23] There are also
differences in the nature of the host responses. When
introduced at high concentrations into tobacco leaves,
incompatible bacteria do not multiply. Daub and Hagedorn,[23]
however, found that in both resistant and susceptible bean
leaves, P. syringae multiplied in the intercellular spaces
and usually was not enveloped. Only rarely was there a
response that resembled envelopment and this occurred with
all host-pathogen combinations. Other studies with susceptible and resistant bean leaves also did not show close attachment between P. phaseolicola and the cell wall.[24] However, in pods there was evidence in all cases of envelopment
of bacteria by fibrillar material from host cell walls.[23]
The response did not prevent bacterial multiplication,
however. This apparent non-specificity of attachment is
consistent with the other studies with bean previously
described.[22]

It is important to examine the possibility that there
is non-specific physical binding in tobacco. One of the

points made by Hildebrand et al.[22] is that films and deposits form very non-specifically, around available objects
such as bacteria and in the interstices between plant cells.
From this, it would be expected that any object infiltrated
into the intercellular spaces of leaves would be expected
to be enveloped. However, when infiltrated into tobacco
leaves in the same manner as bacteria, both polystyrene
balls and asbestos fibers settled on the wall surface upon
drying of the water, but there was no apparent attachment
or envelopment (de Zoeten and Sequeira, unpublished).
Tobacco rattle virus, infiltrated by the same methods, will
attach end-on to tobacco cell walls, but there is no envelopment.[25] If physical non-specific trapping is responsible for envelopment, this should be seen with virus
particles, also. In tobacco, some bacteria do become
trapped in the interstices between host cells. However,
bacteria also are attached along surfaces of cell walls
that do not offer this same potential for trapping. Huang
and Van Dyke[26] provided evidence of immobilization of bacteria by callus cell walls. An incompatible bacterium, P.
pisi, formed aggregates which were enveloped by a network
of fibrillae of plant origin.

In tobacco, true envelopment, rather than physical
entrapment, occurs only with avirulent and incompatible
bacteria. It is argued by Hildebrand et al.[22] that a compatible bacterium, through a combination of extracellular
polysaccharides (EPS) and the surrounding film, can maintain a moist environment and can continue to multiply,
eventually displacing the film and spreading. This could
provide an explanation for the binding of the non-fluidal
avirulent P. solanacearum strains, but not of the fluidal
virulent strains. However, this does not explain why incompatible Race 2 strains of P. solanacearum, which are
attached and induce the HR, also produce large amounts
of EPS.[19]

Hildebrand et al.[22] imply that in the initial stages
of entrapment there is no active host response, but merely
physical effects brought on by changes in the status of
water in intercellular spaces. However, the evidence from
tobacco-bacteria interactions supports the concept of an
active host response. Politis and Goodman[27] followed the
changes in the host cell wall very soon after inoculation
with incompatible P. pisi. As seen in other systems, the

bacteria were in close association with the cell wall after
2 hours. At this time, the plasmalemma had separated from
the cell wall and loose microfibrils began to accumulate.
By 6 hours, organized structures, called appositions, in-
vaginated the plasmalemma opposite the attached bacteria.
Membrane-bound vesicles, that originated from the plasmalemma opposite the bacteria, seemed to carry and deposit
the fibrillar material in the appositions. Their results,
as do the other studies with tobacco, point to an active
host response only in the area close to the point of bacterial attachment, resulting in the formation of specific
structures. This kind of response would not be expected
if the interaction was dependent on mere physical trapping.

     In an electron microscopic study, it is obviously
important to be aware of possible artifacts and not to
interpret physical effects as physiological phenomena.
However, the arguments put forth by Hildebrand et al.[22] do
not invalidate conclusions that apply to the interactions
of bacteria and tobacco. Though physical trapping of bacteria in the interstices between host cells may be responsible for immobilizing some bacteria, it cannot account for
the degree of specificity seen in the attachment process.

Xanthomonas and Erwinia

     Results similar to those found in tobacco-Pseudomonas
interactions have been reported for the cotton-Xanthomonas
malvacearum interaction. Cason et al.[28] found that inoculation of incompatible bacteria into cotyledons of resistant cotton plants caused a detachment of the surface
cuticle and envelopment of adjacent bacteria. The enveloping structures contained fibrillar material. In
contrast, when bacteria were inoculated into susceptible
host tissue, none of these changes in the plant cell wall
occurred and bacteria were not enveloped. Though the
authors felt that the envelopment was too fragile to result
in immobilization of bacteria, they also suggested that the
disruption of the cell wall stimulated by the presence of
the bacteria may be essential for the HR.

     In the interaction of Xanthomonas phaseoli var.
sojensis with soybean, a virulent strain remained free and
multiplied in the intercellular fluid, but two avirulent
strains, which did not induce the HR, were attached to the

mesophyll cell walls and enveloped by fibrous material.
Another strain which elicits the HR was not enveloped,
however.[29] In this case, the results with soybean were
similar to those reached by Sing and Schroth with a bean
system.[21]

The HR can be prevented by maintaining fluid continuously in the intercellular spaces or by introducing bacteria in agar.[30,1] Water-soaking may prevent the HR by
preventing close contact of the bacterium with the host
cell wall, as Stall and Cook have shown for Xanthomonas
vesicatoria introduced into pepper leaves.[1] However, this
may not be the reason for prevention of the HR, because
in some systems the HR is induced without attachment.

When virulent strains of the soft rot bacterium
Erwinia chrysanthemi are introduced into corn leaves, they
remain free and actively dividing in the intercellular
spaces. However, when incompatible E. carotovora strains
are infiltrated, the bacteria become attached and enveloped
by granular material by 4 hours.[31] The plasmalemma separates
slightly from the cell walls and vesicles appear near the
points of bacterial attachment. There is also noticeable
disruption of the bacterial cytoplasm and, in some cells,
apparent detachment of the bacterial envelope. Although
very limited work on the ultrastructure of this interaction has been done, the available evidence suggests that
it follows the pattern described for tobacco and cotton.

THE BACTERIAL COMPONENT OF RECOGNITION

When bacteria invade a potential host, components on
the outer surface of the bacterial cell are the first to
interact with the host cell surface. The high specificity
in these initial interactions suggests that a bacterial cell
surface component is likely to be an important determinant
of specificity. Strong candidates for this role are the
extracellular polysaccharide (EPS) and the lipopolysaccharide (LPS) of the outer membrane. In bacteria with a
true capsule, the oligosaccharide constituents would make
first contact with the plant cell wall and may have sufficient structural stability to bind to specific receptors.
In bacteria with either soluble or no EPS, the LPS would
be a more likely candidate for recognition. In animal
pathogens, the O-antigen of the LPS is highly variable and

determines the antigenic specificity. The lipid A and core regions of LPS are quite constant and would not be expected to vary much within a species.

Agrobacterium tumefaciens

By means of inhibition of tumor formation in the pinto bean bioassay, Whatley et al.[32] studied the nature of the A. tumefaciens component involved in host recognition. Lipopolysaccharides (LPS), extracted from virulent and binding avirulent strains of A. tumefaciens by the standard phenol water method, were highly inhibitory in this assay. A concentration of 1.0 ng/ml gave 35% inhibition of tumor formation. As would be expected of the binding component, inhibition only occurred when LPS was added before or along with the virulent inoculum, but not when added 15 minutes later. The LPS from non-binding strains of A. radiobacter did not show any inhibitory effect. Removal of LPS from bacteria by EDTA treatment reduced tumor-inducing ability of the bacteria.[32]

When Agrobacterium LPS is hydrolyzed by mild acetic acid treatment, the lipid A and the polysaccharide components separate, but only the polysaccharide inhibits tumor formation.[5] Lipid A, when complexed with bovine serum albumin to increase solubility, has no effect on tumor formation. The polysaccharide, which contains the core and O-antigen, showed approximately the same level of biological activity as whole LPS. It is not known what part of the polysaccharide is responsible for recognition, though the O-antigen is a likely candidate because of its greater variability.

The Ti or tumor-inducing plasmid of A. tumefaciens carries sufficient genetic information to transform an avirulent, non-binding strain of A. radiobacter into a virulent, binding strain.[33] When any of several Ti plasmids is introduced into these non-binding strains, their LPS becomes inhibitory in the binding assay; similarly, loss of the plasmid results in loss of binding ability. However, when other virulent strains are cured of their plasmids, they retain site-binding ability, indicating that genetic information for binding may also be located on the chromosome.

Inhibition of gametophore induction in the moss assay was used to determine the bacterial component involved in recognition. Whatley and Spiess[34] demonstrated that LPS of A. tumefaciens is the binding component involved in this interaction. The LPS showed the same binding specificity as whole cells and showed an order of addition effect; LPS was no longer inhibitory when added 24 hours after the virulent cells. As with tumor induction, it is the poly-saccharide, not the lipid A, which is effective as an inhibitor.

Rhizobia-legume interactions

There has been a great deal of debate concerning the identity of the recognition component of Rhizobia. Both EPS and LPS have been proposed. Wolpert and Albersheim[35] examined the interaction between the LPS of four different rhizobia and four different lectins immobilized on columns. Only the LPS of the appropiate symbiont bound to the lectin of its host. Chemical modification of the LPS molecules indicated that the O-antigen portion was active in the binding process. Work by Maier and Brill[36] indicated that the O-antigen may play a role in nodulation of soybeans by R. japonicum. Using paper chromatography of LPS, they found three major differences in composition between the O-antigen of a non-nodulating mutant and that of the wild-type. In another study, it was reported that host plant lectins bound specifically to R. japonicum and R. legumino-sarum and this interaction could be inhibited by LPS extracted from these cells.[37]

There is other evidence the LPS may not be involved in certain Rhizobial interactions. In a study of the LPS from R. leguminosarum, R. phaseoli, and R. trifolii, there was no obvious correlation between immunochemistry or chemical composition of the LPS and the nodulating group.[38] Though the authors did not rule out involvement of LPS, they felt that the data did not lend support to the LPS hypothesis.

Calvert et al.[39] found that ferritin-labeled soybean lectin binds specifically to the capsular (EPS) material of R. japonicum cells and not to the outer membrane (LPS) layer. The ability of R. japonicum to bind soybean lectin changes with different stages in the growth cycle[40] and

these changes are associated with specific changes in the
sugar composition of the capsule.[41] A mutant of R.
leguminosarum which secretes less than normal amounts of
extracellular polysaccharide was incapable of nodulating
host roots, though the parental and mutant types both had
the same LPS structure.[42] This suggests that EPS rather
than LPS may play a role in Rhizobial specificity, though
there may be other explanations for the failure to nodulate
host roots.

In the R. trifolii-clover system, Dazzo and Hubbell[43]
reported that the apparent binding component is an acidic
heteropolysaccharide. Serologically, it is similar to an
antigen present on the surface of clover roots. Since
2-keto-3-deoxyoctonic acid was absent and there was no
endotoxin activity, the authors assumed their preparation
contained no LPS. However, in a more recent paper, Dazzo
and Brill[44] found that the active EPS is serologically
identical to the O-antigen of R. trifolii LPS. R. leguminosarum produces a polysaccharide that is different from
either EPS or LPS.[45] It was suggested that this polysaccharide might help explain the apprent involvement of both
EPS and LPS in recognition. This polysaccharide does not
appear to be part of the LPS.[46]

Other researchers, in trying to resolve this controversy, have attempted to find a role for both LPS and EPS.
One suggestion is that the binding process may actually be
in two steps: an initial recognition involving EPS and a
tighter binding, necessary for furter infection, involving
LPS. Kamberger[47] suggests that EPS is responsible for
absorption of Rhizobium to pea root hairs, but that EPS-
mediated binding is not sufficient for nodulation. A more
specific recognition involving LPS may be necessary for
bacterial invasiveness. In the binding of labeled peanut
lectin to Rizobium, it was reported that there are major
binding sites in EPS, but that there are other binding
sites in LPS.[48]

Pseudomonas solanacearum

All virulent strains of Pseudomonas solanacearum produce large amounts of slime, a soluble, non-capsular form
of EPS. Avirulent, cultural variants of this bacterium
can induce the HR in tobacco or potato leaves and lack EPS.

Sequeira and Graham[49] found that a low concentration of potato lectin agglutinated avirulent, but not virulent cells of the bacterium. If virulent cells were washed to remove EPS, they were agglutinated to a certain degree, but when EPS was added back, agglutination was reduced. The LPS from either HR or non-HR inducing bacteria was precipitated by the lectin, though the latter not as extensively. It appeared from these data that LPS is the binding component in both HR and non-HR inducing strains, but that in virulent strains, EPS prevents binding and allows bacterial multiplication.[49] Whether the effect of EPS was due to competitive binding or a general blocking effect was not established.

Further studies with a series of virulent and avirulent strains, the latter including HR and non-HR inducers, indicated a strong correlation between LPS structure and ability to induce the HR.[50] Based on sugar composition of LPS, particularly on the relative amounts of xylose and glucose, and on mobility by SDS polyacrylamide gel electrophoresis, it was shown that all the HR-inducing strains had rough (R) LPS, that is, they lacked part or all of the O-antigen.[50] All but one of nine non-HR inducing strains had smooth (S) or complete LPS. The one exception (strain Q) which had R-LPS based on gel separation and was intermediate in sugar composition, may have been unable to induce the HR for other reasons. Strain Q is an intermediate strain that will be useful to define the precise configuration of LPS that is necessary for binding to host cell wall. A bacteriophage that lyses both virulent and avirulent non-HR inducers, but not the HR inducers, is also being used to establish LPS structure in these strains. Pretreatment of the phage with S-LPS inhibited lysis of a virulent strain of the bacterium, whereas R-LPS had no effect.[51]

These studies suggested that the differences in LPS structure determine the ability of P. solanacearum to cause the HR. Also, they suggest that the binding component resides in either the core or lipid A portion, since O-antigen is lacking in the rough strains. In the smooth LPS, the highly branched O-antigen side-chain may mask the recognition site.

Lipopolysaccharide also plays a role in induced resistance in tobacco leaves. If heat-killed avirulent or

virulent Pseudomonas cells are introduced into tobacco
leaves, the leaves will be protected against subsequent chal-
lenge by a variety of pathogens.[52,53] In protected leaves,
induction of the HR by avirulent P. solanacearum is prevented. Heat-killed cells attach and are enveloped in a
manner similar to live HR-inducing cells,[19] but no HR results.
Graham et al.[54] found that protection was induced by LPS
alone, whether from rough or smooth strains, and determined
that the active portion was located in the core-lipid A
region. When LPS was injected into leaves, ultrastructural
changes occurred in host cells similar to those obtained
with heat-killed cells. The role of LPS in HR induction
and in induced resistance evidently is not the same, since
the former requires rough LPS, while the latter is obtained
with either R- or S-LPS. Similarly, Mazzuchi et al.[55,56]
found that both the HR and susceptible response in tobacco
could be prevented by a protein-LPS complex from phytopathogenic pseudomonads.

Xanthomonas and Erwinia

In the interaction of Xanthomonas phaseoli var.
sojensis with soybean, it was shown that virulent strains
do not bind to host walls, while avirulent do bind and are
enveloped.[29] All five strains examined had equal amounts
of EPS of identical sugar composition. The virulent strains,
however, all had S-LPS, as determined by sugar composition
and SDS polyacrylamide gel electrophoresis. In contrast,
the avirulent all had predominantly R-LPS. LPS from virulent strains had approximately 30% fucose and less than 10%
glucose, while the avirulent contained no detectable fucose
and over 30% glucose (Whatley, unpublished). The difference
in binding between virulent and avirulent would appear,
therefore, to reside in the LPS. In other Xanthomonas
systems, however, there is some evidence that the EPS plays
a role in determining specificity.[57]

It is possible that the specificity of other bac-
terial-plant interactions depends on the rough-smooth difference in LPS structure. The O-antigen may provide a way
to mask the recognition component and allow the pathogen to
remain undetected. It is also possible that an indirect
effect may be involved. There can be major changes in cell
membrane protein associated with smooth-rough transitions[58]
which may be important. For example, Bruegger and Keen

found a specific elicitor of glyceollin production in soybean which could be extracted from the cell envelope of P. glycinea.[59] Neither LPS nor EPS were active as elicitors; the elictor appears to be a glycoprotein.

Strains of Erwinia stewartii, which are virulent on corn, produce large amounts of EPS. These cells, as well as several avirulent, EPS-producers, were not agglutinated by a corn agglutinin. Those avirulent strains which lack EPS were readily agglutinated.[60] There were no differences between the LPS from agglutinable and non-agglutinable strains (Whatley, unpublished). It appears that LPS may be involved in binding, while EPS may prevent the interaction of the LPS with the plant receptor.

THE PLANT CELL WALL RECEPTOR

In most instances, the bacterial component recognized by the host appears to be a polysaccharide, either LPS or EPS. Therefore, the plant cell wall receptor must be able to recognize polysaccharides specifically. Because lectins are proteins or glycoproteins which are able to bind specific sugars, these proteins fit the requirements for receptors. Lectins can be located on host cell walls and there is some evidence that they play a role in plant-bacterial interactions, although there is still no definitive proof. To obtain proof, it is important to show that the lectin binds the bacterial polysaccharide specifically in vivo and in vitro, that this binding is hapten-reversible, and that the lectin is localized in the parts of the plant that normally bind the bacteria. Other possible receptors must be considered. For instance, a plant polysaccharide could interact specifically with the bacterial polysaccharide.[57,61,62] Evidence points in the direction of a carbohydrate-carbohydrate interaction in a few systems.

Agrobacterium tumefaciens

Since a wound is necessary for tumor induction, it is possible that the plant binding component could be on an exposed plasma membrane or on the cell wall. By means of the pinto bean bioassay, Lippincott et al.[63] concluded that the binding component resides on the cell wall. In the tumor induction assay, the inhibitory effect of the cell wall preparations could be neutralized by pretreatment of

the cell walls with certain whole heat-killed bacteria or
their LPS. Non-binding cells or their LPS do not alter
the binding ability of the cell walls, indicating that the
same binding specificity is exhibited in vivo and in vitro.

Cell wall preparations from a number of different
plants or tissue cultures were tested for their binding
ability to inhibit tumor formation.[64] Cell walls from
dicots, but not those from monocots were inhibitory.[64]
Though this finding correlates with the host range of
Agrobacterium, the difference in binding probably does not
explain the specificity of the interactions. In the same
study, it was indicated that after transformation, cell
walls from tumors are no longer able to bind Agrobacterium,
indicating that there are changes in cell wall structure
associated with tumor induction.

When isolated cell wall components were tested, it
was found that polygalacturonic acid was highly inhibitory
and pectin was mildly inhibitory.[62] Removal of pectin and
hemicelluloses by strong acid treatment reduced the binding
ability of the cell walls. Differences in effectiveness of
pectin and polygalacturonic acid appeared to be due to differences in degree of methylation. When non-binding cell
walls, such as those from monocots, tumors, or embryonic
tissue, were pretreated with pectin methylesterase, they
became highly inhibitory. Lippincott and Lippincott concluded that the binding site may involve the polygalacturonic acid in the middle lamella and primary cell wall.[62]
In its highly methylated form (pectin), polygalacturonic
acid loses this binding ability. In spite of this evidence,
the possibility that lectins are involved in recognition
has not been ruled out. Because A. tumefaciens has such a
wide host range, it is likely that the binding component
is similarly non-specific, such as polygalacturonic acid.
Recognition in crown gall tumor induction may involve an
LPS-polygalacturonic acid interaction.

Rhizobium-legume interactions

The role of lectins in Rhizobium-legume interactions
has been reviewed recently[65,66] and will be mentioned here
only briefly. Hamblin and Kent[67] first suggested a role
for lectin in recognition with the demonstration that there
was a specific binding of bean lectin to R. phaseoli.

Bohlool and Schmidt[68] showed later that soybean lectin (SBL) combined specifically with 22 of 25 nodulating strains of R. japonicum. They did not find binding of the lectin to non-nodulating strains. The SBL binding sites are polar,[69] a fact which agrees with the end-on attachment reported by Dazzo et al.[17] These results of Bohlool and Schmidt were supported by extensive tests in Bauer's laboratory and the important criterion of hapten specificity was met, since N-acetyl-D-galactosamine, an SBL hapten, fully blocked bacterial-lectin binding specifically.[40]

There remained, however, the problems of nodulating strains that did not bind the lectin and of the non-nodulating ones that did.[15] The first problem was partly solved by the demonstration that many of these non-binding bacteria, when grown in soybean root exudates, developed receptors for SBL.[70] Discrepancies in results, therefore, may be due to differences in the culture conditions used. It also must be remembered that specific binding to the lectin need not always lead to nodulation, since there are many steps in the process at which it could be blocked.

There are many other unresolved questions. For example, a specific hapten for SBL is apparently not present in R. japonicum EPS,[71] though binding sites in situ do not necessarily have to involve a simple hapten. Also, SBL has not been detected in the roots of soybean at an age when the plant could still be nodulated.[72] An immunological study indicated that the roots of a number of soybean lines appeared to lack SBL totally, though SBL was found in a line thought to lack it.[73] There are a number of possible explanations for this. For example, lectin may be more tightly bound and not extractable by standard procedures.[72] It is necessary to settle these questions before a role for lectin can be demonstrated in the R. japonicum-soybean interaction.

The R. trifolii-clover system also has been extensively studied. Dazzo and Hubbell[43] reported that a carbohydrate antigen on the surface of clover roots was cross-reactive with a bacterial polysaccharide. A lectin named trifoliin was isolated from clover seed and was found to bind to both the bacteria and clover roots.[74,75] Binding of bacteria by trifoliin can be inhibited by the sugar 2-deoxy-D-glucose, a compound which can also elute lectin

from clover seedling roots. It was proposed that trifol-
liin forms a bridge between bacterium and host to establish
binding.[43,66] The major problem with this hyopthesis is
that there is little evidence that trifoliin is on the roots.

Pseudomonas solanacearum

The bacterial component involved in the induction of
the hypersensitive response (HR) in tobacco by Pseudomonas
solanacearum seems to be LPS. Whether EPS plays a role in
preventing binding of bacteria in vivo is still an open
question. A lectin isolated from potato or tobacco agglutinated the HR-inducing strains of P. solanacearum much more
strongly than virulent strains, and this was correlated
with a stronger precipitation of R-LPS compared with S-LPS.[49]
Fluorescein-labeled lectin bound to avirulent, but not to
virulent cells of P. solanacearum and this binding was
inhibited by chitin oligomers, which contain internal
N-acetyl glucosamine moieties, the hapten for potato
lectin.[49] Potato and tobacco lectins, therefore, seem to
meet the requirement of specificity of binding of the bacterial recognition component. A nitrocellulose filter
binding assay is currently being used to quantitate differences in binding between LPS and EPS of different
strains.[76]

It is also important to show the lectin is present in
tissues which bind bacteria. A lectin, with agglutinating
specificity identical to that of the potato lectin, was
isolated from tobacco leaves by infiltrating the leaves
with saline and recovering the intercellular fluid by
centrifugation.[49] That the lectin was located on the cell
walls was demonstrated with the use of fluorescent antilectin antiserum.[77] Binding of the labeled antibodies to
mesophyll cell walls of tobacco and potato was demonstrated
by fluorescence microscopy. The potato and tobacco lectins
are hydroxy-proline-rich proteins, as are other plant cell
wall components, and also contain arabinogalactans, substances commonly found in the cell wall.[49] In further
studies of localization, it will be necessary, therefore,
to establish that the antiserum is specific for the lectin
and not for other cell wall components.

Although most of the evidence points to the involvement of the lectin in the plant-bacterial interaction

resulting in the HR, further evidence is necessary in terms of localization of the lectin, hapten reversibility, and quantitation of binding of bacterial cell wall components.

Xanthomonas and Erwinia

A possible role for plant agglutinins in soybean-Xanthomonas phaseoli var. sojensis interactions was examined by Fett.[29] Standard SBL did not agglutinate the bacterium, but a new soybean agglutinin differentially agglutinated different strains of the bacterium. However, there was no apparent correlation with virulence, although these strains differed in LPS structure and in their ability to bind to cell walls. Both EPS and LPS inhibited agglutination by the agglutinin. It appears that either a lectin is not involved in this interaction or the correct one has not been identified as yet.

On the basis of in vitro interactions, Morris et al.[57] suggested that Xanthomonas-plant interactions may involve polysaccharides of both organisms. Xanthans, EPS isolated from Xanthomonas spp., gel when mixed with plant galactomannans, even at low concentrations. They suggest that this interaction may provide a "molecular holdfast" for the bacterium on the plant.[57] It is also reported that there is specificity in the gelling of both plant and bacterial components.[78] Though the in vitro studies show a specific interaction, there is as yet no in vivo demonstration of the role of this polysaccharide-polysaccharide interaction as a determinant of specificity.

An agglutinin extracted from corn seed agglutinated the avirulent strains of E. stewartii which did not produce EPS. Virulent strains, all of which produce EPS, were not agglutinated.[60] A role for this agglutinin in E. stewartii-corn interactions remains to be demonstrated, however.

CONCLUDING REMARKS

In the systems that have been studied, there is strong physical and chemical evidence for specific site-attachment between bacteria and plants. In two of these systems, root nodule formation and crown gall tumor induction, the binding

is necessary for the continued interaction of the bacteria
with the host. In those interactions that result in an incompatible, hypersensitive response, attachment appears to
be part of a host defense mechanism. There are, however,
unexplained cases of HR induction without prior attachment.
In either compatible or incompatible interactions, attachment results in a host response. How this response is
linked to attachment is unknown, but it is one of the key
questions in the entire area of recognition. There must
be a system for transfer of information from the site of
attachment on the cell wall to the target organelle in the
host cytoplasm.

In all cases, the bacterial component appears to be
either a cell surface polysaccharide (LPS or EPS) or a
glycoprotein. However, the identity of the bacterial component in most interactions, including the Rhizobium-legume
interaction, is still in doubt. There has been a great
deal of conflicting data generated by researchers who have
used different systems. An attempt must be made to study
the same bacterial strains and the same host cultivars in
different laboratories if the question is to be resolved.

The complementary plant component involved in recognition may be either a lectin or a polysaccharide. The interaction of bacteria and plant cell wall components must be
sufficiently specific to explain the highly predictable
nature of the host responses. However, the role of lectins
in plant bacterial binding has not been adequately demonstrated and there is little evidence for polysaccharide-
polysaccharide interactions. Future studies on lectins
must center on the localization of these substances in the
appropriate parts of the plant and on the specificity and
hapten reversibility of their interaction. It is important
to move past the stage of simple correlations into a detailed examination of chemical structure of both interacting
components and the direct demonstration of their involvement
in recognition. In particular, structural modifications
which prevent the interactions should be examined.

These questions regarding the identity of the interacting components must be settled before we can begin to
examine how changes in both host and parasite are stimulated after attachment.

REFERENCES

1. Stall, R. E. and A. A. Cook. 1979. Evidence that
    bacterial contact with the plant cell is necessary for
    the hypersensitive reaction but not the susceptible
    reaction. Physiol. Plant Pathol. 14:77-84.
2. Lippincott, B. B. and J. A. Lippincott. 1969. Bac-
    terial attachment to a specific wound site as an es-
    sential stage in tumor initiation by Agrobacterium
    tumefaciens. J. Bacteriol. 97:620-628.
3. Schilperoort, R. A. 1969. Investigations on plant
    tumors. Crown gall: On the biochemistry of tumor-
    induction by Agrobacterium tumefaciens. Thesis,
    University of Leiden.
4. Beiderbeck, R. 1973. Bacterial cell wall and tumor
    induction by Agrobacterium tumefaciens. Z.
    Naturforschung. Part C. 28:19-201.
5. Whatley, M. H. 1977. Studies on the adherence step
    exxential for tumor induction by Agrobacterium.
    Thesis, Northwestern University, Evanston, IL.
6. Glogowski, W. and A. G. Galsky. 1978. Agrobacterium
    tumefaciens site attachment as necessary prerequisite
    for crown gall tumor formation on potato discs.
    Plant Physiol. 61:1031-1033.
7. Bogers, R. J. 1972. On the interaction of A.
    tumefaciens with cells of Kalanchoe daigremontiana.
    In Proc. 3rd Int. Conf. Plant Pathogenic Bacteria,
    (H. P. Maas Geesteranus, ed.). Wageningen, the Nether-
    lands: Cent. Agri. Publ. Doc. pp. 239-250.
8. Matthysse, A. G. and P. M. Wyman. 1978. Attachment
    of Agrobacterium tumefaciens to tissue culture cells.
    Plant Physiol. 61S:72.
9. Ohyama, K., L. E. Pelcher, A. Schaefer, and L. C. Fowke.
    1979. In vitro binding of Agrobacterium tumefaciens
    to plant cells from suspension culture. Plant Physiol.
    63:382-387.
10. Chilton, M.-D., M. H. Drummond, D. J. Merlo, D. Sciaky,
    A. L. Montoya, M. P. Gordon and E. W. Nester. 1977.
    Stable incorporation of plasmid DNA into higher plant
    cells: The molecular basis of crown gall tumorigenesis.
    Cell 11:263-271.
11. Drummond, M. H., M. P. Gordon, E. W. Nester, and M.-D.
    Chilton. 1977. Foreign DNA of bacterial plasmid
    origin is transcribed in crown gall tumors. Nature
    269:535-536.

12. Spiess, L. D., B. B. Lippincott, and J. A. Lippincott.
     1971. Development and gametophore induction in the
     moss Pylaisiella selwynii as influenced by Agrobac-
     terium tumefaciens. Amer. J. Bot. 58:726-731.
13. Spiess, L. D., B. B. Lippincott, and J. A. Lippincott.
     1976. The requirement of physical contact for moss
     gametophore induction by Agrobacterium tumefaciens.
     Amer. J. Bot. 63:324-328.
14. Spiess, L. D., J. C. Turner, P. G. Mahlberg, B. B.
     Lippincott, and J. A. Lippincott. 1977. Adherence
     of agrobacteria to moss protonema and gametophores
     viewed by scanning electron microscopy. Amer. J.
     Bot. 64:1200-1208.
15. Chen, A.-P. T. and D. A. Phillips. 1976. Attachment
     of Rhizobium to legume roots as the basis for specific
     interactions. Plant Physiol. 38:83-88.
16. Napoli, C. A., F. B. Dazzo, and D. H. Hubbell. 1975.
     Production of cellulose microfibrils by Rhizobium.
     Appl. Microbiol. 30:123-131.
17. Dazzo, F. B., C. A. Napoli, and D. H. Hubbell. 1976.
     Adsorption of bacteria to roots as related to host
     specificity in the Rhizobium-clover symbiosis. Appl.
     Environ. Microbiol. 32:166-171.
18. Klement, Z. 1963. Method for the rapid detection
     of pathogenicity of phytopathogenic pseudomonads.
     Nature 199:299-300.
19. Sequeira, L., G. Gaard, and G. A. de Zoeten. 1977.
     Attachment of bacteria to host cell walls: Its rela-
     tion to mechanisms of induced resistance. Physiol.
     Plant Pathol. 10:43-50.
20. Goodman, R. N., P. Y. Huang, and J. A. White. 1976.
     Ultrastructural evidence for immobilization of an
     incompatible bacterium, Pseudomonas pisi, in tobacco
     leaf tissue. Phytopathology 66:754-764.
21. Sing, V. O., and M. N. Schroth. 1977. Bacteria-plant
     cell surface interactions: Active immobilization of
     saprophytic bacteria in plant leaves. Science
     197:759-761.
22. Hildebrand, D. C., M.-C. Alosi, and M. N. Schroth.
     1980. Physical entrapment of pseudomonads in bean
     leaves by films formed at air-water interfaces.
     Phytopathology 70:98-109.

23. Daub, M. E. and D. J. Hagedorn. 1980. Growth kinetics
     and interactions of Pseudomonas syringae with suscep-
     tible and resistant bean tissues. Phytopathology
     70:429-436.
24. Sigee, D. C., and H. A. S. Epton. 1975. Ultrastruc-
     ture of Pseudomonas phaseolicola in resistant and
     susceptible leaves of French bean. Physiol. Plant
     Pathol. 6:29-34.
25. Gaard, G. and G. A. de Zoeten. 1979. Plant virus
     uncoating as a result of virus-cell wall interactions.
     Virology 96:21-31.
26. Huang, J. S. and G. C. Van Dyke. 1978. Interaction
     of tobacco callus tissue with Pseudomonas tabaci, P.
     pisi, and P. fluorescens. Physiol. Plant Pathol.
     13:65-72.
27. Politis, D. J. and R. N. Goodman. 1978. Localized
     cell wall appositions: Incompatibility response of
     tobacco leaf cells to Pseudomonas pisi. Phytopathology
     68:309-316.
28. Cason, E. T., Jr., P. E. Richardson, M. K. Essenberg,
     L. A. Brinkerhoff, W. M. Johnson, and R. J. Venere.
     1978. Ultrastructural cell wall alterations in immune
     cotton leaves inoculated with Xanthomonas malvacearum.
     Phytopathology 68:1015-1021.
29. Fett, W. F. 1979. Occurrence and physiological
     properties of Pseudomonas glycinea and Xanthomonas
     phaseoli var. sojensis in Wisconsin and presence of a
     bacterial agglutinating factor in soybean. Thesis,
     University of Wisconsin, Madison, WI.
30. Cook, A. A. and R. E. Stall. 1977. Effects of water-
     soaking on response to Xanthomonas vesicatoria in
     pepper leaves. Phytopathology 67:1101-1103.
31. Victoria, J. I. 1977. Resistance in corn (Zea mays
     L.) to bacterial stalk rot in relation to virulence
     of strains of Erwinia chrysanthemi. Thesis, Univer-
     sity of Wisconsin, Madison, WI.
32. Whatley, M. H., J. S. Bodwin, B. B. Lippincott, and
     J. A. Lippincott. 1976. Role for Agrobacterium cell
     envelope lipopolysaccharide in infection site attach-
     ment. Infect. Immun. 13:1080-1083.
33. Whatley, M. H., J. B. Margot, J. Schell, B. B.
     Lippincott, and J. A. Lippincott. 1978. Plasmid and
     chromosomal determination of Agrobacterium adherence
     specificity. J. Gen. Microbiol. 107:395-398.

34. Whatley, M. H. and L. D. Spiess. 1977. Role of
     bacterial lipopolysaccharide in attachment of Agro-
     bacterium to moss. Plant Physiol. 60:765-766.
35. Wolpert, J. S. and P. Albersheim. 1976. Host-
     symbiont interactions. I. The lectins of legumes
     interact specifically with the O-antigen containing
     lipopolysaccharide of their symbiont rhizobia.
     Biochem. Biophys. Res. Comm. 70:729-737.
36. Maier, R. J. and W. J. Brill. 1978. Involvement of
     Rhizobium japonicum O-antigen in soybean nodulation.
     J. Bacteriol. 133:1295-1299.
37. Kato, G., Y. Maruyama, and M. Nakamura. 1979. Role
     of lectins and lipopolysaccharide in the recognition
     process of specific legume-Rhizobium symbiosis.
     Agric. Biol. Chem. 43:1085-1092.
38. Carlson, R. W., R. E. Sanders, C. Napoli, and P.
     Albersheim. 1978. Host-symbiont interactions. III.
     Purification and partial characterization of Rhizobium
     lipopolysaccharides. Plant Physiol. 62:912-917.
39. Calvert, H. E., M. Lalonde, T. V. Bhuvaneswari, and
     W. D. Bauer. 1978. Role of lectins in plant-micro-
     organism interactions. IV. Ultrastructural local-
     ization of soybean lectin binding sites on Rhizobium
     japonicum. Can. J. Microbiol. 24:785-793.
40. Bhuvaneswari, T. V., S. G. Pueppke, and W. D. Bauer.
     1977. Role of lectins in plant-microorganism inter-
     actions. I. Binding of soybean lectins to rhizobia.
     Plant Physiol. 60:486-491.
41. Mort, A. and W. D. Bauer. 1978. The chemical basis
     of lectin binding to Rhizobium japonicum. Plant
     Physiol. 61S:59.
42. Sanders, R. E., R. W. Carlson, and P. Albersheim.
     1978. A Rhizobium mutant incapable of nodulation and
     normal polysaccharide secretion. Nature 271:240-242.
43. Dazzo, F. B. and D. H. Hubbell. 1975. Cross-reactive
     antigens and lectin as determinants of symbiotic spe-
     cificity in the Rhizobium-clover association. Appl.
     Microbiol. 30:1017-1033.
44. Dazzo, F. B. and W. J. Brill. 1979. Bacterial poly-
     saccharide which binds Rhizobium trifolii to clover
     root hairs. J. Bacteriol. 137:1362-1373.
45. Planqué, K. and W. J. Kijne. 1977. Binding of pea
     lectins to a glycan type polysaccharide in the cell
     walls of Rhizobium leguminosarum. FEBS Lett.
     73:64-66.

46. Planqué, K., J. J. Nierop, and A. Burgers. 1979.
The lipopolysaccharide of free-living and bacteroid
forms of Rhizobium leguminosarum. J. Gen. Microbiol.
110:151-159.
47. Kamberger, W. 1979. Role of cell surface polysaccharides in the Rhizobium-pea symbiosis. FEMS
Microbiol. Lett. 6:361-365.
48. Bhagwat, A. A. and J. Thomas. 1980. Dual binding
sites for peanut lectin on Rhizobia. J. Gen.
Microbiol. 117:119-125.
49. Sequeira, L. and T. L. Graham. 1977. Agglutination
of avirulent strains of Pseudomonas solanacearum by
potato lectin. Physiol. Plant Pathol. 11:43-54.
50. Whatley, M. H., N. Hunter, M. A. Cantrell, C. A.
Hendrick, K. Keegstra, and L. Sequeira. 1980. Specific changes in Pseudomonas lipopolysaccharide
associated with induction of the hypersensitive
response in tobacco. Plant Physiol. 65:557-559.
51. Hendrick, C. A., M. H. Whatley, N. Hunter, M. A.
Cantrell, and L. Sequeira. 1979. The hypersensitive
response in tobacco: A phage capable of differentia-
ting HR and non-HR-inducing Pseudomonas solanacearum.
Plant Physiol. 63S:134.
52. Lovrekovich, L. and G. L. Farkas. 1965. Induced
protection against wildfire disease in tobacco leaves
treated with heat-killed bacteria. Nature (London)
205:823-824.
53. Lozano, J. C. and L. Sequeira. 1970. Prevention of
the hypersensitive reaction in tobacco leaves by heat-
killed bacterial cells. Phytopathology 60:875-879.
54. Graham, T. L., L. Sequeira, and T.-S. R. Huang. 1977.
Bacterial lipopolysaccharides as inducers of disease
resistance in tobacco. Appl. Environ. Microbiol.
34:424-432.
55. Mazzuchi, U. and P. Pupillo. 1976. Prevention of
confluent hypersensitive necrosis in tobacco by a
bacterial protein-lipopolysaccharide complex.
Physiol. Plant Pathol. 9:101-112.
56. Mazzuchi, U., C. Bazzi, and P. Pupillo. 1979. The
inhibition of susceptible and hypersensitive reactions by protein-lipopolysaccharide complexes from
phytopathogenic pseudomonads: Relationship to polysaccharide antigenic determinants. Physiol. Plant
Pathol. 14:19-30.

57. Morris, E. R., D. A. Rees, G. Young, M. D. Walkinshaw,
     and A. Darke. 1977. Order-disorder transition for a
     bacterial polysaccharide in solution. A role for
     polysaccharide conformation in recognition between
     Xanthomonas pathogen and its host plant. J. Molec.
     Biol. 110:1-16.
58. Gmeiner, J. and S. Schlecht. 1979. Molecular organ-
     ization of the outer membrane of Salmonella typhimurium.
     Eur. J. Biochem. 93:609-620.
59. Bruegger, B. B. and N. T. Keen. 1979. Specific
     elicitors of glyceollin accumulation in the Pseu-
     domonas glycinea-soybean host-parasite system.
     Physiol. Plant Pathol. 15:43-51.
60. Bradshaw-Rouse, J., L. Sequeira, A. Kelman, and D.
     Coplin. 1980. Extracellular polysaccharide and
     virulence of Erwinia stewartii in relation to aggluti-
     nation by a corn lectin. Phytopathology 71: (In
     press).
61. Moorhouse, R., W. T. Winter, and S. Arnott. 1977.
     Conformation and molecular organization in fibers
     of the capsular polysaccharide from E. coli M41
     mutant. J. Molec. Biol. 109:373-391.
62. Lippincott, J. A. and B. B. Lippincott. 1977.
     Nature and specificity of the bacterium-host attach-
     ment in Agrobacterium infection. In Cell Wall Bio-
     chemistry Related to Specificity in Host-Plant
     Pathogen Interactions, (B. Solheim and J. Raa, eds.).
     Norway Universitetsforlaget, Oslo. pp. 439-451.
63. Lippincott, B. B., M. H. Whatley, and J. A. Lippincott.
     1977. Tumor induction by Agrobacterium involves at-
     tachment to a site on the host plant cell wall.
     Plant Physiol. 59:388-390.
64. Lippincott, J. A. and B. B. Lippincott. 1978. Cell
     walls of crown-gall tumors and embryonic plant tis-
     sues lack Agrobacterium adherence sites. Science
     199:1075-1078.
65. Sequeira, L. 1978. Lectins and their role in host-
     pathogen specificity. Annu. Rev. Phytopathol.
     16:453-481.
66. Dazzo, F. B. 1980. Adsorption of microorganisms to
     roots and other plant surfaces. In Adsorption of
     Microorganisms to Surfaces, (G. Bitton and K. C.
     Marshall, eds.). John Wiley and Sons, Inc.
     pp. 253-316.

67. Hamblin, J. and S. P. Kent. 1973. Possible role of phytohaemagglutinin in Phaseolus vulgaris L. Nature 245:28-30.
68. Bohlool, B. B. and E. L. Schmidt. 1974. Lectins: A possible basis for specificity in the Rhizobium-legume root nodule symbiosis. Science 185:269-271.
69. Bohlool, B. B. and E. L. Schmidt. 1976. Immuno-fluorescent polar tips of Rhizobium japonicum: Possible site of attachment of lectin binding. J. Bacteriol. 125:118-194.
70. Bhuvaneswari, T. V. and W. D. Bauer. 1978. Role of lectins in plant-microorganism interactions. III. The influence of rhizosphere/rhizoplane culture conditions on the soybean lectin-binding properties of rhizobia. Plant Physiol. 62:71-74.
71. Mort, A. J., M. E. Slodki, R. D. Plattner, and W. D. Bauer. 1979. The initiation of infections in soybean by Rhizobium. 4. Molecular structure of biologically active R. japonicum polysaccharides. Plant Physiol. 63S:135.
72. Pueppke, S. G., W. D. Bauer, K. Keegstra, and A. L. Ferguson. 1978. Role of lectins in plant-microorganism interactions. II. Distribution of soybean lectin in tissues of Glycine max (L.) Merr. Plant Physiol. 61:779-784.
73. Su, L.-C., S. G. Pueppke, and H. P. Friedman. 1980. Lectins and the soybean-Rhizobium symbiosis. I. Immunological investigations of soybean lines, the seeds of which have been reported to lack the 120,000 dalton soybean lectin. Biochem. Biophys. Acta 629:292-304.
74. Dazzo, F. B. and W. J. Brill. 1977. Receptor site on clover and alfalfa roots for Rhizobium. Appl. Environ. Microbiol. 33:132-136.
75. Dazzo, F. B., W. E. Yanke, and W. J. Brill. 1978. Trifoliin: A Rhizobium recognition protein from white clover. Biochim. Biophys. Acta 529:276-286.
76. Duvick, J. P., L. Sequeira, and T. L. Graham. 1979. Binding of Pseudomonas solanacearum suface polysaccharides to plant lectin in vitro. Plant Physiol. 63S:134.
77. Leach, J., M. A. Cantrell, and L. Sequeira. 1978. Localization of potato lectin by means of fluorescent antibody techniques. Phytopathol. News 12:197.

78. Dea, I. C. M., E. R. Morris, D. A. Rees, E. J. Welsh,
     H. A. Barnes, and J. Price. 1977. Associations of
     like and unlike polysaccharides: Mechanism and spe-
     cificity in galactomannans, interacting bacterial
     polysaccharides, and related systems. Carbohydr.
     Res. 57:249-272.

Chapter Ten

LECTINS AND PLANT-HERBIVORE INTERACTIONS

DANIEL H. JANZEN

Department of Biology
University of Pennsylvania
Philadelphia, Pennsylvania   19104

INTRODUCTION

When an animal takes a bite out of a plant, it is
gustatorily and digestively treading on a battlefield im-
planted and strewn with traits generated by natural selection during millions of years of acts of herbivory. The
glycoproteins called lectins are heterogeneous in kind,
place and density in this battlefield. How animals respond
to this pattern and its parts suggest that lectins may be
more than simply one more of nature's many kinds of glue.
They are quite unfortunately called "lectins" since it is
their sticky nature that is the basis of their biological
function; while they are specific in their attachment to
certain sugars, the fact that the same sugars can occur on
the surface of many different kinds of cells makes them in
fact highly variable in the specificity of their stickiness
at the cellular level of organization (which surely is the
level of concern when one calls them "lectins").

What are the relevant properties of a glue?
    How tightly does it adhere to two or more
        relevant surfaces?
    How specific is its adherence?
    How quickly and easily can it be dissolved or
        neutralized by the user?
    How expensive is it?

How can it be used for other purposes?
What is its shelf life?
How much is needed to do the job?
How long does it retain its properties once in
  place?
What is its setting time?

If we ask these sorts of questions of a child's paste
pot, of a welder's rod, of electrician's tape, of a gar-
dener's tanglefoot, of epoxy resins, of a spider's web, of
a flower's stigma, to say nothing of what Fort Dix thought
up during the Viet Nam war, we quickly note that lectins
have a number of properties which should lead, through
natural selection, to their becoming on occasion part of
a plant's defense repertoire. Just as there are specific
places in battle where boiling oil is an important direct
defense (though hot oil is generally much more important
as a motor lubricant), a moderate-sized protein that
adheres tightly to very specific sorts of surfaces can on
occasion be an effective defensive weapon.[1] Needless to
say, direct defense against other organisms is certainly
not the biological function of all lectins any more than
providing a microbe binding site on roots is the biological
function of the lectins in cecropia moth blood.[2]

Among the array of lectin kinds, concentrations and
placements, where is it reasonable to search for a defense
function against multicellular herbivores? Before embarking
on this query, I must very briefly mention what lectins
apparently do to a herbivore that consumes them. They apparently bind to gut surfaces, be they of the animal or of
its essential microbial fauna, and alter the function of
those surfaces.[3] In short, they fall in the category of
digestion inhibitors, as do protease inhibitors,[4,5] poly-
phenols (tannin, lignin, etc.),[5,6] cellulose,[7] unassimilable
starches, etc. Since a lectin molecule is small relative
to the herbivore's gut surface area, and since it has only
a few active sites,[3] for a lectin to be a significant
digestion inhibitor it will have to occur as a relatively
large proportion of the diet.

Without dipping further into the philosophical background of why study protein glues as defensive compounds
as well as molecules useful to the plant in other ways, I
would like to relate the early unfoldings of a study of an

exceptional little mouse, a study that bears strongly on lectins as defenses against seed predators.

LIOMYS SALVINI

Liomys salvini (Figure 1) is a hispid pocket mouse in the same family of professional seed-eaters, Heteromyidae, as is the more familiar desert kangaroo rat, Dipodomys. L. salvini, or 'guarda fiesta' as it is locally known, however is a forest-floor mouse. This tropical mouse is a common inhabitant of the deciduous forests of Guanacaste Province, on the lowland coastal plain of Costa Rica.[8-10] Its diet is almost entirely seeds of herbs, vines, shrubs and trees, which it finds in the litter and in animal dung. It also shells seeds out of certain species of newly fallen fruits. The specific area where I have studied this mouse, Santa Rosa National Park (about 35 km north of Liberia, Guanacaste Province, Costa Rica), has a flora of about 650 species of broad-leafed plants,[11] over half of which are woody perennials with seeds large enough for this mouse to bother with. Of this set of seeds, it is conspicuous that certain species

Figure 1. Adult female Liomys salvini (46 g) from Santa Rosa National Park, Guanacaste Province, Costa Rica (July, 1980).

are flatly rejected (Table 1). These seeds lie about on
the forest floor, in or out of their fruits, for months
without being eaten. They are never encountered in the
pouches of trapped mice. They were rejected when presented
to the mice as part of seed mixes on plates at which free-
ranging mice nightly foraged in the forest.

It is possible that in some absolute sense some
species of seeds are rejected because they are nutritionally a badly imbalanced food; however, this does not explain why they are not eaten as, for example, a carbohydrate source in combinations with other seed species.
Some are rejected because they are in a container too hard
or unwieldy for L. salvini to gnaw through (Acrocomia
vinifera palm nuts are an example), but the great majority
of rejected seeds are no harder than those that are eaten
and are often bitten into in apparent testing behavior.
I find the most reasonable working hypothesis to be that
these seeds are rejected for the chemicals that they contain other than those that are normally thought of as
dietary nutrients for small mammals. Since this essay is
directed at the obnoxious protein portion of a seed's
defenses, I will focus on that aspect rather than some of
the other known (additional) potential defenses of these
seeds (e.g., Mucuna pruriens seeds contain L-dopa,[12]

Table 1. Some species of native forest and forest-edge woody
plant seeds and nuts eagerly eaten and thoroughly rejected by
free-foraging Liomys salvini in Santa Rosa National Park.

Rejected	Accepted
Acrocomia vinifera	Phaseolus lunatus
Spondias mombin	Enterolobium cyclocarpum
Lonchocarpus acuminatus	Cochlospermum vitifolium
Lonchocarpus costaricensis	Malvaviscus arboreus
Pithecellobium platylobum	Hymenaea courbaril
Canavalia brasiliensis	Forsteronia spicata
Canavalia maritima	Pithecellobium saman
Dioclea megacarpa	Cissus rhombifolia
Mucuna pruriens	Sesbania emerus
Ateleia herbert-smithii	
Cassia grandis	

Dioclea and Canavalia seeds contain canavanine,[13] Ateleia
seeds contain 2,4-methanoproline and 2,4-methanoglutamic
acid,[14] etc).

With the current understanding of potentially toxic
seed proteins, protease inhibitors and lectins are the two
protein-aceous candidates to examine with respect to Liomys
rejection of seeds as food. Turning briefly to protease
inhibitors, in the accepted column of Table 1 there is
Enterolobium cyclocarpum, a mimosaceous legume seed that
contains a substantial amount of a protease inhibitor
(C. A. Ryan, personal communication), but no lectins, at
least as measured by agglutination of human, rabbit or
hamster blood (I. Liener, personal communication). This
seed, in the hard (dormant, ungerminated) or germinated
state can serve as the sole diet of L. salvini for a
month or more in the laboratory,[10] is avidly sought by L.
salvini in the field, and is chosen over many other accep-
table seed species in choice tests. This suggests that L.
salvini may be sufficiently specialized as a seed predator
that it carries gut enzymes or other digestive repertoires
generally resistant to protease inhibitors (it is unlikely,
however, to be like the bruchid beetle larvae that are
resistant to protease inhibitors by having a digestive
system that operates largely without proteases[15]). To
test this, I fed 4 or 5 L. salvini on diets of pure rat
chow, and on rat chow that was 1, 5 or 25% soybean typsin
inhibitor (Table 2). There was no reduction in weight (and
even a hint of weight increase compared to animals on pure
rat chow) or rejection of the adulterated food, nor was
there any visible symptom of a toxic or otherwise debil-
itating effect of the adulterated food. As a working hypothesis, I will assume that L. salvini is highly resistant
to protease inhibitors in legume seeds.

There are at least 4 species of seeds in the rejected
column of Table 1 that show strong agglutinating activity
towards human, rabbit and hamster red blood cells (I. Liener,
personal communication): Canavalia brasiliensis, C.
maritima, Dioclea megacarpa, and Cassia grandis. The most
desirable test would be to extract the lectins from these
seeds and feed them to L. salvini. This step is planned.
However, black beans (Phaseolus vulgaris) are a rich source
of a potent lectin that has been shown to be lethal when
incorporated in the diet of the larvae of a bruchid beetle.[1]

Table 2. Weight changes and survival of adult <u>Liomys</u> <u>salvini</u> (hispid pocket mice) fed a variety of diets containing potentially toxic proteins. The animals were wild-caught and maintained in individual cages with <u>ad</u> <u>lib</u> food and water at temperatures approximating those of their natural habitat (Santa Rosa National Park, Guanacaste Province, Costa Rica, May-July, 1980.

Treatment	Mouse number	Initial[1] weight (g)	Weight change (g)	Final Health	Duration of feeding (days)	Weight change per day (g)
Rat chow:						
	37 ♂	32	+1	healthy	9	+0.11
	38 ♀	40	+1	healthy	7	+0.14
	39 ♀	41	-4	healthy	7	-0.57
	40 ♂	59	0	healthy	7	0
	41 ♀	36	+1	eaten by boa	4	+0.25
X̄		41.6	-0.2			-0.02
s.d.		10.4	2.2			0.32
Rat chow with 1% soybean trypsin inhibitor:						
	1 ♀	50	0	healthy	6	0
	5 ♀	40	0	healthy	6	0
	7 ♀	41	-1	healthy	6	-0.17
	8 ♀	40	0	healthy	6	0
	10 ♀	47	+1	healthy	6	+0.17
X̄		43.6	0.0			0.00
s.d.		4.6	0.7			0.12
Rat chow with 5% soybean trypsin inhibitor:						
	3 ♀	45	+1	healthy	6	+0.17
	9 ♀	43	+2	healthy	6	+0.17
	12 ♀	43	+2	healthy	6	+0.33
	15 ♀	38	+1	healthy	6	+0.17
X̄		42.3	+1.3			+0.21
s.d.		3.0	0.5			0.08
Rat chow with 25% soybean trypsin inhibitor:						
	37 ♂	33	0	healthy	5	0
	38 ♀	41	+1	healthy	5	+0.20
	39 ♀	37	+1	healthy	5	+0.20
	40 ♂	59	+1	healthy	5	+0.20
X̄		42.5	+0.8			+0.15
s.d.		11.5	0.5			+0.10

Table 2 (continued)

Treatment	Mouse number	Initial[1] weight (g)	Weight change (g)	Final Health	Duration of feeding (days)	Weight change per day (g)
Cooked black beans:						
	51 ♀	¨	+1	healthy	6	+0.17
	53 ♀	45	-1	healthy	5	-0.20
	54 ♀	25	+4	healthy	5	+0.80
	58 ♀	31	+1	healthy	5	+0.20
	59 ♂	54	-3	healthy	5	-0.60
X̄		38.8	+0.4			+0.07
s.d.		11.4	2.6			0.52
Uncooked black beans:						
	1 ♀	46	-7	dying[2]	3	-2.33
	3 ♀	36	-7	dying[2]	3	-2.33
	5 ♀	41	-6	dying[2]	3	-2.00
	7 ♀	35	-4	dying[2]	3	-1.33
	50 ♀	44	-12	dead	10	-1.20
	52 ♀	38	-6	dead	3	-2.00
	55 ♀	40	-3	dead	3	-1.00
	56 ♀	34	-5	dead	3	-1.67
	57 ♀	35	-4	dead	3	-1.33
X̄		38.8	-6.0			-1.69
s.d.		4.3	2.6			0.50
Animals that refused to eat toxic seeds:[3]						
	24 ♀	48	-15	dead	7	-2.14
	21 ♂	36	-9	dead	4	-2.25
	27 ♀	46	-10	dead	6	-1.67
	28 ♂	38	-10	dead	7	-1.43
	29 ♀	38	-9	dead	4	-2.25
	30 ♀	32	-6	dead	2	-3.00
X̄		39.7	-9.8			-2.12
s.d.		6.1	2.9			0.55

[1] When a mouse appears more than once in this table, the first time its "initial weight" is the weight at the time of capture. The second time its "initial weight" is the weight in the laboratory following at least 5 days in the laboratory feeding on high quality food.

[2] These 4 mice would not have lived another 12 hours and had the usual traits of starving mice in their last few hours of life (severe shakes, closing eyes, poor coordination, little response to stimulation).

[3] Given the data available, I will assume that mice that voluntarily starve themselves to death die at the same rate as those deprived of all food.

This result is particularly striking because bruchids as a group are seed predators.[16] Black beans are cheap and easily available in Costa Rica. They also suffer no depredation by rats when stored in rat infested habitations.

When L. salvini were offered a diet consisting solely of
black beans that had been boiled for one hour, they maintained their body weight and other wise appeared quite
healthy (Table 2). However, when given a pure diet of
only uncooked black beans, L. salvini lost weight at a rate
no different from that of those eating no food at all
(Table 2; $t_{13 \, d.f}$ = 1.54, n.s). The one animal out of 9
that still appeared healthy after the third day survived
for 10 days and probably was in exceptionally good condition at the beginning of the experiment. During the course
of the experiment, the mice did eat small but highly variable
amounts of uncooked black bean seeds. The mouse that lived
so long (number 50 ♀, Table 2) ate roughly half the weight
of uncooked black bean seeds per day as it would have were
it maintaining its body weight on laboratory rat chow. It
lost 27% of its body weight before death, which is nearly
twice the loss these mice usually tolerate.

As a working hypothesis, I conclude that it is the
lectin in the black beans that is killing the mice. The
mode of action is probably the combined effect of direct
starvation caused by food rejection (averaged over 34 mouse
days, the bean consumption per day per mouse was 0.16 gram,
and these mice require 2 to 4 grams per day of laboratory
rat chow to maintain their body weight) and reduced nutrient
uptake through the intestinal wall, as suggested by there
being no conspicuous relationship between rate of weight
loss and amount of bean eaten. While there are protease
inhibitors in black beans, the impressive ability of L.
salvini to live on food rich in soybean trypsin inhibitor
and live on protease inhibitor-rich Enterolobium cyclocarpum
seeds suggests that protease inhibitors are not the likely
cause of black bean seed rejection by L. salvini. There
are no known alkaloids or uncommon amino acids in commercial
black beans, nor are there any other known potentially
toxic molecules besides the proteins. The ultimate test
of this working hypothesis depends on the availability of
purified black bean and other lectins in 10 to 30 gram
amounts. L. salvini is a very abundant seed predator in
the forests it occupies. If it can be definitively shown
that it cannot eat seeds rich in lectins, it can be stated
with certainty that these seed lectins serve as a defense
role against rodents irrespective of their other uses to
the seedling. Lectins in the diet at naturally occurring
concentrations can kill the larvae of a seed predator

bruchid.[1] This demonstrates that seed lectins are functional as are alkaloids, uncommon amino acids, cyanogenic
glycosides, etc. in defending seeds against insect seed
predators.

If a dietary chemical is lethal to L. salvini, the
situation is more than the mere act of hitting a naive gut
with just any potentially nasty compound. This seed-eating
specialist is unaffected by the very protease inhibitors
that are a major reason for humans to boil or otherwise
process seeds before eating them. For example, the seeds
of E. cyclocarpum are lethal if they are the sole diet of
Sigmodon hispidus, another terrestrial rodent in the same
habitat (though S. hispidus does quite well on them if
boiled).[17] L. salvini can live on a pure diet of HCN-rich
Phaseolus lunatus seeds. E. cyclocarpum seeds are also
rich in pipecolic acid and albizzine (E. A. Bell, personal
communication), two uncommon amino acids with conspicuous
insecticidal properties at the concentrations found in
seeds.[18,19] Sesbania emerus seeds are rich in canavanine
(G. A. Rosenthal, personal communication) yet eaten readily
by L. salvini. This animal has a versatile gut yet there
appears to be at least one lectin that it cannot handle.

WHAT PROCESSES EVOLUTIONARILY PUT LECTINS IN SEED CHEMICAL
REPERTOIRES?

There are five traits of the system that are relevant
this question:

1) The forest has many kinds of seeds, and the seeds
of each species contain a unique combination of potentially
defensive compounds.

2) The forest has many kinds of seed predators, each
with the ability to ignore or detoxify some of these compounds, but not all.

3) Lectins are just one of the many protective traits
a seed contains.

4) There are two different seed-predator responses
to a seed, each likely to generate different traits in
the defense array.

a) The rodent-type animals, such as <u>Liomys</u> <u>salvini</u>, try a newly encountered seed, and if is has the appropriate defenses, reject it. Rejection will depend as well on hunger, gut conditioning, body weight of the animal, alternate available foods, health and fat condition of the animal, perceptability of the defense compound, etc. The selective pressure favoring better-defended mutants is essentially constant over the years. This is because each year there are new recruits in the habitat that have to learn about the seeds of that habitat and season, and because there is a rodent-specific rate of forgetting that leads to re-sampling and relearning as each species of plant comes into seed again year after year. Since small rodents are very common, an unprotected mutant will be quickly located and its seed crop probably eliminated by the local set of mice. On the other hand, as selection for resistance traits occurs, the seed will have to be a consistently important part of the diet of the mouse for there to be such a strong selection that the mouse evolutionarily increases or shifts its detoxification abilities to encompass the change. The more usual rodent response to a mutant seed that is better defended should be to eat less of it. This system does not proceed to total inedibility for all seeds because the mice are also dispersal agents, defenses have economic costs to both parent and off-spring, plants and rodent densities fluctuate, rodent detoxification abilities do change, etc.

b) The insect-type animals oviposit on a particular species of host seed (or on its fruit), usually the only species of seed in the habitat that their larvae can eat.[16] Their larvae are specialists on the chemisty of that seed.[20-22] In addition, all mature and immature stages are behaviorally as well as physiologically programmed to deal with the host plant's other traits (e.g., timing of fruiting, fruit chemistry, odor cues for location, susceptibility to parasitoid attack while in that species of seed, etc.). While these animals may be very regular and deadly in killing their host's seeds, they pose no threat in contemporary time to nearly all other plant species present in the habitat, even if the larvae could develop in their seeds. There are undoubtedly many

other seed species in the habitat, which, when only
seed chemistry is considered, could serve as hosts
for a particular species of insect that does not, in
fact, feed on them.

However, probing of other hosts does occur in evolutionary time, but rarely. A probe occurs when an ovi-
positing female makes an oviposition error or owing to a
shortage of its regular hosts, oviposits on almost anything that contains even a fraction of the oviposition
stimuli emitted by the usual host. Futhermore, in the
latter case (and possibly the former) a successful probe
may lead to a shift in the host seed species used, rather
than a broadening of the host list. This is because it is
likely that the beetle will have a higher fitness as a
specialist on either one or the other hosts than as a
generalist on both, since thorough bypassing of a plants
defenses often requires very fine tuning at the behavioral,
morphological, physiological, biochemical, etc. level.
Fine tuning may well be impossible owing to differences in
timing of seed production, seed chemistry, fruit morphology,
etc. If L. salvini or small rodents cannot handle lectins
in general, this will select for strong convergence in
lectin traits in the seeds these rodents confront; convergence will be on that molecule that works the best at an
optimal cost for the genetic lineage of the parent and
offspring that bear it.

It is the insect-type seed predator that should be
responsible for much of the fine tuning of a lectin's
traits, when that lectin is serving primarily as a defense
compound in a seed. If each kind of lectin requires a different chemistry of detoxification by a host-specific insect, then a seed's lectin traits will be occasionally
evolutionarily modified because the mutant repels a host-
specific insect seed predator, just as is the case with
alkaloids, uncommon amino acids, cyanogenic glycosides,
etc. In contrast to the case where a group of seed predators (e.g., rodents) can bypass a class of compounds (e.g.,
Liomys versus protease inhibitors) or is repelled by a
class of compounds (e.g., rodents versus lectins), each
time a lectin's properties are changed it is a novel defense as seen by the host-specific insect. Here, then,
natural selection will not result in convergence of traits
among lectins, but rather in the continual appearance of

new types. There should even be active selection that
results in divergence in lectin types because a mutant,
that is a change in a direction already occupied by other
lectins, is likely to become susceptible to the host-
specific seed predators that can bypass those other lectins.
Both diversification and divergence in lectin types should,
however, approach an equilibrium level. The level should
be determined in part by all those other ecological processes besides physiological seed availability that determine the numbers, kinds and diversity of seed predators in
the habitat, and in part by the other non-predator-related
selective pressures on lectins.

5) Like other compounds found in the seedling's bag
lunch, lectins should be under strong selection to be of
multiple use in this weight-, volume-, resource quality-
limited container. The ideal combination of compounds in
a seed is that which maximizes the fitness of that genetic
lineage. Surely this will require a complex balance of

a) partitioning of parental resources among the seeds
    (seed size, weight, number, etc.),

b) seed photosynthesis and therefore contribution to
    its own resources, and

c) seed resources for seedling growth and protection
    against herbivores (probably no molecule is ideal
    for both functions).

Specifically, lectins will be the focus of selective
pressures associated with

a) their use as glue in development, such as in the
    attachment of symbiotic bacteria to roots or in
    the binding of different cells within the organism,

b) their degradative destruction as an amino acid and
    small polypeptide source in seedling metabolism,

c) their use as protective compounds in the cotyledons
    and in newly produced vegetative tissues (deter-
    rentsto contemporary herbivory by generalists and
    evolution of herbivory by specialists).

In some plants, past selection will have been such that lectins were never functional as anything more than one kind of glue, a glue of importance only in some very internal biochemical sense. Here, then, other compounds are the defenses and storage compounds in the seeds. But somewhere early in the dim history of legume seeds, there was a combination of herbivore susceptibility and plant lectin synthesis capacity that led to lectin-rich seeds becoming an integral part of the character of a 'successful' species of plant which then radiated in various ways to give us many species of lectin-rich legume seeds. Alternatively, one may hypothesize that this even occurred many times, owing to the general presence of the lectin-synthesis ability of legumes, which in turn pre-disposed them to selection by herbivores that got bad stomach aches from diets rich in protein glues. One cannot chose between these two historical scenarios with the data at hand, but it is obvious what sort of amino acid sequence studies are needed to distinguish among them. They are also not mutually incompatible. Owing to the chemical complexity of proteins, as contrasted with small defensive molecules like alkaloids and uncommon amino acids which can have absolutely convergent end products in their synthesis by different plants, the history of a protein molecule is to some more extreme degreee incorporated in its structure. The question has become not 'Is a lectin for defense?', but rather 'What are the various ways that sticky glycoproteins are functional?' 'What selects for their detailed traits?' and 'What selects for deposition in certain plant parts in sufficient bulk to give a herbivore a gut ache?', bearing in mind that herbivores come in widely differing sizes, food consumption rates and intensity of desire to eat a particular plant part.

WHY ARE LECTINS SO PROMINENT IN SEEDS AND TUBERS?

Assuming that the relatively high concentrations of kinds and amounts of lectins in seeds and tubers is not a sampling artifact, there are several plausible ecological reasons why this severe distributional heterogeneity should occur.

Dilution. Many species of animals that commit severe seed predation on mature seeds are sufficiently specialized on this diet that they eat almost no other food for all or

much of their lives. This means that the contents of the
seed are likely to be all or nearly all of their stomach
and intestinal contents at a given time. That is to say,
whatever is in the seed runs little risk of being diluted
out in a larger bulk of different food (as occurs, for
example, when a horse digests some of the hard lectin-rich
legume seeds it swallows along with leafy feeds). On the
other hand, seed contents are very concentrated nutrients,
and compared to a foliage-eating animal, a seed-eater consumes a miniscule amount of food. For example, a bruchid
beetle larva in a legume seed may only consume twice its
last instar body volume in seed contents during its entire
development, while a moth caterpillar eating leaves may
consume is own volume of food during every 24 hours of
active growth. In short, the amount of lectin required
to maintain a 5% lectin titer in a seed-eater's gut is
easily only 1 to 5% of that which would be required for
the same effect in a foliage-eater's gut of the same body
weight. That is to say, an expensive digestion inhibitor
like a lectin or protease inhibitor may be economically
most appropriate for a seed while the much cheaper (per
gram) polyphenol digestion inhibitors are most appropriate
for foliage. While both may be found in each type of plant
part, here I am discussing the forces that keep the dispro-
portionalities in the system.

   Bag lunch. Seeds and tubers are storage devices, and
in the case of the former, volume- and weight-conscious
ones. A lectin (as well as a protease inhibitor) may
double as a polypeptide and amino acid storage unit (and
of course may even have its evolutionary roots in an innocuous storage protein). However, to the degree that only
innocuous storage proteins are found in more vegetative plant
parts, we are then again left with the disproportionality
question. Futhermore, while it is all very well to beat
your sword into a ploughshare, it may be best to put it
into the closet and buy a plough when the war is over,
because wars have a way of reappearing each generation.
Finally, to give you a very high quality sword, your mother
may well have used such high quality steel that it makes a
lousy ploughshare.

   Fitness. Finally, and most definitely not least,
there is the simple answer that gram for gram, seeds and
tubers have the highest fitness value of any plant part.

From the viewpoint of the juvenile in the seed, the seed
is the essence of fitness. For the parent plant, the
seeds are one of its throws of the dice to remain in a
surviving lineage. Seeds are probably the most thoroughly
protected of all plant parts and part of that thoroughness
is achieved by containing quite physiologically active
compounds of many kinds in high concentrations. Virtually
all seeds contain at least 1 potent digestion inhibitor
and 2 to 5 small molecules (cyanogenic glycosides, alkaloids, cardiac glycosides, uncommon amino acids, cyano-
lipids, etc.)[23] unless they are involved in some sort of
population-level seed predator satiation (e.g., as in
conifers, oaks, bamboos, niloo, chestnuts, dipterocarps,
etc.),[24-26] physical protection (hard nuts), or very small
size (many herbs). Our own village histories undoubtedly
taught us this. Seeds and tubers are where the goodies
are, but you cannot eat them unless either you process
them (cook them, break them, dig them up, breed out their
defenses) or they are so chemically defenseless as to be
eaten by the bulk of the herbivorous animals in the habitat.
The same ecological story applies to the presence of protease inhibitors and largely indigestible starches and
other complex sugars in seeds and tubers, but I shall let
that lie as it is not the subject of this symposium. On
the other hand, it may be noted that the intensity of protection required is related not only to the value to the
owner (mother and seed), but the value to the thief.
Seeds and tubers contain the highest concentrations of
animal nutrients in the plant world, and often occur at a
density in time and space quite high enough to support many
species of herbivores for much if not all of the year or
generation. Such an array of barbarian hordes cannot be
kept at bay with a few ditches and spears.

ACKNOWLEDGMENTS

This paper was supported by NSF DEB 80-11558 and was
written in response to invitation to this conference. It
has profited from constructive editorial comments by W.
Hallwachs, and conversations with I. Liener and C. A. Ryan.
It has its conceptual origin in the realization that toxic
but heat-denaturable proteins in seeds were probably the
impetus to human invention of cooking in general, or at
the least to the boiling of seeds.

REFERENCES

1. Janzen, D. H., H. B. Juster and I. E. Liener. 1976.
    Insecticidal action of the phytohemagglutinin in black
    beans on a bruchid beetle. Science 192:795-796.
2. Yeaton, R. L. W. 1980. Lectins of a North American
    silkmoth (Hyalophora cecropia): their molecular char-
    acterization and developmental biology. Ph.D. thesis,
    University of Pennsylvania, Philadelphia, 229 pp.
3. Liener, I. E. 1979. Phytohemagglutnins. In Herbi-
    vores, Their Interactions with Secondary Plant Meta-
    bolites. (G. A. Rosenthal and D. H. Janzen, eds.).
    Academic Press, New York, pp. 567-598.
4. Ryan, C. A. 1979. Proteinase inhibitors. In
    Herbivores, Their Interactions with Secondary Plant
    Metabolites. (G. A. Rosenthal and D. H. Janzen, eds.).
    Academic Press, New York, pp. 599-618.
5. Feeny, P. 1976. Plant apparency and chemical defense.
    Recent Adv. Phytochem. 10:1-40.
6. Rhoades, D. F. and R. G. Cates. 1976. Toward a
    general theory of plant antiherbivore chemistry.
    Recent Adv. Phytochem. 10:168-213.
7. Janzen, D. H. 1979. New horizons in the biology of
    plant defenses. In Herbivores, Their Interactions
    with Secondary Plant Metabolites. (G. A. Rosenthal
    and D. H. Janzen, eds.). Academic Press, New York,
    pp. 331-350.
8. Fleming, T. H. 1974. The population ecology of two
    species of Costa Rican heteromyid rodents. Ecology
    55:493-510.
9. Bonoff, M. B. and D. H. Janzen. 1980. Small terres-
    trial rodents in 11 habitats in Santa Rosa National
    Park, Costa Rica. Brenesia 17:163-173.
10. Hallwachs, W. and D. H. Janzen. 1981. Adequacy of
    Enterolobium cyclocarpum seeds as diet for Liomys
    salvini. J. Mammal. (submitted).
11. Janzen, D. H. and R. Liesner. 1980. Annotated check-
    list of plants of lowland Guanacaste Province,
    Costa Rica, exclusive of grasses and non-vascular
    cryptogams. Brenesia 18:15-90.
12. Bell, E. A. and D. H. Janzen. 1971. Medical and
    ecological considerations of L-DOPA and 5-HTP in seeds.
    Nature 229:136-137.

13. Rosenthal, G. A. 1977. The biological effects and mode of action of L-canavanine, a structural analogue of L-arginine. Quart. Rev. Biol. 52:155-178.
14. Bell, E. A., M. Y. Qureshi, R. J. Pryce, D. H. Janzen, P. Lemke and J. Clardy. 1980. 2,4-Methanoproline (2-carboxy-2,4-methanopyrrolidine) and 2,4-methanoglutamic acid (1-amino-1,3-dicarboxycyclobutane) in seeds of Ateleia herbert-smithii Pittier (Leguminosae). J. Amer. Chem. Soc. 102:1409-1412.
15. Applebaum, S. W. 1964. Physiological aspects of host specificity in the Bruchidae - I. General considerations of developmental compatibility. J. Insect Physiol. 10:783-788.
16. Janzen, D. H. 1980. Specificity of seed-attacking beetles in a Costa Rican deciduous forest. J. Ecol. 68:929-952.
17. Hallwachs, W. and D. H. Janzen. 1981. Toxicity of Enterolobium cyclocarpum seeds to Costa Rican Sigmodon hispidus. J. Mammal. (submitted).
18. Rehr, S. S., E. A. Bell, D. H. Janzen, and P. Feeny. 1973. Insecticidal amino acids in legume seeds. Biochemical Syst. 1:63-64.
19. Janzen, D. H., H. B. Juster and E. A. Bell. 1977. Toxicity of secondary compounds to the seed-eating larvae of the bruchid beetle Callosobruchus maculatus. Phytochemisty 16:223-227.
20. Rosenthal, G. A., D. L. Dahlman and D. H. Janzen. 1976. A novel means for detailing with L-canavanine, a toxic metabolite. Science 192:256-258.
21. Rosenthal, G. A., D. H. Janzen and D. L. Dahlman. 1977. Degradation and detoxification of canavanine by a specialized seed predator. Science 196:658-660.
22. Rosenthal, G. A., D. L. Dahlman and D. H. Janzen. 1978. L-canaline detoxification: a seed predator's biochemical mechanism. Science 202:528-529.
23. Janzen, D. H. 1978. The ecology and evolutionary biology of seed chemistry as relates to seed predation. In Biochemical Aspects of Plant and Animal Coevolutin. (J. B. Harbone, ed.). Academic Press, London, p. 163-206.
24. Janzen, D. H. 1971. Seed predation by animals. Annu. Rev. Ecol. Syst. 2:465-492.
25. Janzen, D. H. 1974. Tropical blackwater rivers, animals, and mast fruiting by the Dipterocarpaceae. Biotropica 6:69-102.

26.   Janzen, D. H.   1976.   Why do bamboos wait so long to flower?   <u>Annu</u>. <u>Rev</u>. <u>Ecol</u>. <u>Syst</u>.   <u>7</u>:347-391.

Chapter Eleven

CELL INTERACTIONS AND PATTERN FORMATION IN DICTYOSTELIUM
DISCOIDEUM

DANIEL McMAHON

Program in Genetics and Department of Zoology
Washington State University
Pullman, Washington 99164

INTRODUCTION

The panoply of forms of plants and animals is as
amazing as is the reproducibility with which each form is
produced within a given species of plant or animal. The
reproducibility is determined by a molecular mechanism
through which cells in the organism determine their relative position. This process leads to the pattern of spots
on the frog, stripes on a tiger, and the precise intercon-
nections which occur between the ganglion cells of retina
and their target neurons in the optic tectum of the frog.[1]
In the cellular slime mold, Dictyostelium discoideum, the
process of cellular position determination occurs in the
pseudoplasmodium and leads to two types of cells: stalk
cells and spore cells. Approximately 1/3 of the cells in
the pseudoplasmodium become stalk cells; most of the remaining two-thirds of the cells become spores.[2] The pro-
portioning is largely invariant over a wide variety of
experimental conditions and over an approximately one
hundred thousand fold difference in number of cells in
the pseudoplasmodium.

While the developmental cycle of D. discoideum is
rapidly becoming common knowledge among biologists, I can
briefly summarize the major cellular events which occur
during development. Unicellular amoebae reproduce asexu-
ally under favorable environmental conditions, including
the presence of ample supplies of food. If the amoebae
begin to starve, they become chemotactic to cyclic AMP and
also begin to emit pulses of cyclic AMP. This behavior
causes groups of amoebae to aggregate together. Their aggregation is facilitated by an increase in intracellular
adhesivity which occurs at the same time. The resulting
aggregate may be composed of up to one hundred thousand
cells. The cells of the aggregate, under appropriate
environmental conditions, form a migrating, slug shaped
pseudoplasmodium. At this stage, the cells of the pseudoplasmodium determine their relative position and use this
information to determine their direction of their differentiation. This fact was elegantly demonstrated by Kenneth
Raper.[3] In his experiments, reciprocal grafts were made
between pseudoplasmodia which had been produced from cells
which had fed on colorless bacteria or on the red bacterium,
Serratia marcescens. If a tip was taken from a red pseudoplasmodium and grafted onto a colorless pseudoplasmodium
from which the tip had been removed, the stalk of the resulting sorocarp was colored red and the spores were uncolored. In the reciprocal graft, spores were colored red,
and the stalk was uncolored. This experiment helped demonstrate that the cells of the pseudoplasmodium become pre-
destined as a function of their position in the pseudoplasmodium.

A substantial number of experiments have shown that
detectable position-dependent biochemical changes occur in
the cells of the pseudoplasmodium. These include histochemical differences between the cells at the front and the
cells at the rear. The content of periodic acid-Schiff's
staining carbohydrates, alkaline phosphatase, and glycogen
phosphorylase differ substantially between cells of the
front and the back.[4,5,6]

THE CELL CONTACT MODEL

A number of molecular theories for the mechanism by
which cells might determine their position in a morphogenetic field, such as the pseudoplasmodium, have been

advanced. These have been reviewed elsewhere.[7] I have proposed a theory which suggests that this mechanism is mediated by complimentary molecular contacts between molecules on the plasma membranes of adjacent cells.[8]

Propositions contained in this model are the following:

1. Cells have complimentary contact-sensing molecules on their cell surface.

2. These molecules, when activated by contact, have complimentary effects on the intracellular concentration of a morphogen, such as cyclic AMP, which regulates metabolism and gene action.

3. The effective "concentration" of these molecules on the cell surface is regulated by negative feedback (i.e., as the concentration of intracellular morphogen (e.g., cAMP) rises within the responding cell, the system which increases cyclic AMP is inactivated, whereas the system which degrades cyclic AMP is activated).

4. These molecules have a polarized distribution on the cell surface.

A schematic diagram which illustrates the operation of this model is presented in Figure 1. In Figure 1, the top row of six boxes represents six cells in a morphogenetic field. Each of the cells is identical both in concentration of intracellular morphogen as denoted by capital A and in their display of cell surface molecules as denoted by F and R. A simple physical fact is responsible for the initiation and propagation of changes in surface display. In each line of cells, the front cell has no cell in front of it and the rear cell has no cell behind it. Assuming that the contact-sensing molecules on the rear of the cell, designated by R in this figure, increase the intracellular concentration of cyclic AMP, and the contact-sensing molecules on the front of the cell, designated by F in this figure, decrease the concentration of cyclic AMP, a propagating change in intracellular cyclic AMP concentration occurs. The first cell in line has R molecules activated by contact with the second cell. Its F is not activated by contact since no cell precedes it. The system which

increases the concentration of cyclic AMP is activated.
The system which degrades it is not. Therefore, the intracellular concentration of cyclic A increases within the
cell.   As a consequence, the number (or activity) of F's
is increased, whereas the number (or activity) or R's is
decreased.   As the process proceeds, the second cell in
line becomes functionally equivalent to the first cell in
line (i.e., the R molecules on the cell preceding it disappear).   Each row of cells in Figure 1 illustrates the
progression of this process with time.   The final result
is a group of cells, as illustrated in the bottom row, in
which the front cells have large intracellular concentrations of A, whereas the rear cells have small intracellular

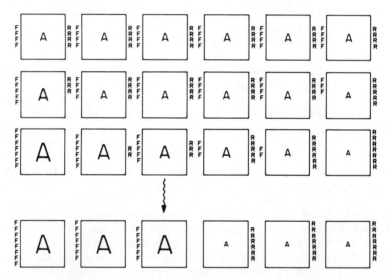

Figure 1.   Schematic description of the cell contact model.
Each row of boxes in this diagram represents a row of cells
in a morphogenetic field.   The letters represent: A, the
concentration of intracellular morphogen; F, the contact-
sensing molecules on the front of a cell; R, the contact-
sensing molecules on the rear of a cell.

concentrations of A. In addition, the front cells have on their surface a substantial number of F molecules and lack R molecules. The reverse is true of the cells in the rear.

This model can be represented by simple differential equations and these equations were solved[8] on the digital computer, numerically, using the Runge-Kutta method. The result of this numerical simulation of this system led to results which agree with those illustrated in Figure 1. It predicted that a large intracellular concentration of cyclic AMP should be present in the cells which are destined to become stalk cells and that a sharp boundary should separate this from the cells destined to become spore cells, which would contain low intracellular concentrations of cyclic AMP.

TESTS OF THE MODEL

A number of predictions arise from the cell contact model. First, the model predicts that a discontinuous gradient of intracellular cyclic AMP should be produced as described above. Second, the model suggested that cyclic AMP was a mediator of the cytodifferentiation of th stalk cell. Third, the model indicates that a continuous chain of normally reacting cells is necessary for the propagation of information in a morphogenetic field. Fourth, the model suggests that intracellular signaling in the morphogenetic field is mediated by molecules found on the plasma membrane of the developing cells. Fifth, it suggested that these contact-sensing molecules should become regionally localized on cells of the pseudoplasmodium as a function of their position.

The first proposition of the model, i.e., that cyclic AMP would show an unusual distribution in cells of the pseudoplasmodium was tested by Pan et al.[9] They used an antiserum which had been raised against cyclic AMP. It was made fluorescent by conjugation with fluorescein isothicyanate. Cells and sections of pseudoplasmodium were stained with the serum. These experiments demonstrated that a discontinuous gradient of cyclic AMP develops in the pseudoplasmodium. As predicted by the model, this histochemical method demonstrates that the cells of the front have high intracellular concentration of cyclic AMP and a sharp discontinuity separates these from the cells

of the rear with much lower concentrations of cyclic AMP.
Experiments by others[10],[11] indicate that there is a 50%
difference in intracellular cyclic AMP that can be detected between the cells of the front and the rear of the
pseudoplasmodium by radioimmunoassay. We (Goldberg and
McMahon, unpublished results) have looked at the content
of cyclic AMP in cells of the stalk and in spore cells at
the time the pseudoplasmodium completes culmination, also
by radioimmuno-assay. The stalk cells contain 10 times as
much cyclic AMP as the spore cells.

Klaus and George[12] have examined the necessity for a
continuous chain of viable cells for the normal propagation
of information in the pseudoplasmodium of D. discoideum.
They irradiated a pseudoplasmodium with a microbeam laser
to kill small groups of cells within the pseudoplasmodium.
When this was done, the pseudoplasmodium frequently fragmented into two pseudoplasmodia. They concluded that the
results of their experiments agreed with the predictions
of the model.

We have examined the ability of purified plasma membranes to interact with living cells during morphogenesis
biochemical differentiation. Plasma membranes were purified by the method of McMahon et al.[13] from cells at various
stages of development. Pseudoplasmodial cells were dissociated by trituration and plated on Millipore filters in
the presence or absence of added plasma membranes. These
experiments yielded a number of interesting results[14]
(McMahon, unpublished results). Morphogenesis of the dissociated cells is prevented by added plasma membranes.
Untreated cells are able to reform pseudoplasmodia within
four hours after being placed on a Millipore filter and
to culminate within six hours. Cells treated with plasma
membrane were able to reaggregate but morphogenesis was
disrupted at this point.

When dissociated pseudoplasmodial cells were plated
on Millipore filters, the developmentally-regulated enzyme,
alkaline phosphatase, increases in activity four hours
after replating the cells; reaches a maximum of activity
after approximately eight hours; then proceeds to decline
in activity. If cells are plated on the filter in the
presence of plasma membranes purified from pseudoplasmodial
cells, this enzyme is superinduced. The superinduced level

of this activity is maintained at least throughout the
next twelve hours.[14]

We investigated several characteristics of this superinduction of alkaline phosphatase activity by membranes.
These experiments provided the following results: first,
when extracts of membranes are mixed with extracts of cells
in vitro, the enzyme is not activated; second, the effect
required contact between the cells and the plasma membranes
(it is not exhibited if the cells are separated from plasma
membranes by a Millipore filter with a pore-size of 0.45
microns); third, the effect is developmentally specific.
If vegetative amoebae are treated with pseudoplasmodial
plasma membranes, the amoebae do not show an increase in
alkaline phosphatase activity. On the other hand, plasma
membranes prepared from vegetative amoebae are several
times less active on a protein basis than pseudoplasmodial
cell membranes in causing the superinduction of alkaline
phosphatase when pseudoplasmodial cells are treated with
them.

Treatment of the membranes with Pronase under conditions which degrade the major polypeptides of the plasma
membrane but have little effect on the size of the glycoproteins do not destroy the activity of the membranes in
this assay, whereas mild periodate oxidation of the plasma
membrane does (McMahon, unpublished results). Since pure
plasma membranes can be isolated which contain little contamination except the small percentage contamination with
the inner membrane of the mitochondria,[13,15] these effects
appeared to be caused by a component of the plasma membrane.

MACROMOLECULAR COMPOSITION OF THE PLASMA MEMBRANE

To examine the chemical and topographical changes
which occur in the macromolecular components of the plasma
membrane during development, we have examined the composition of the plasma membrane on SDS gels at various stages
of development. Vegetative amoebae, aggregating amoebae,
and pseudoplasmodial cells have been compared. The plasma
membrane of the vegetative amoebae contained 55 major polypeptides. 50% of these change amount during development.
In order to determine whether any of the polypeptides of
the external surface of the plasma membrane change in surface exposure during development, cells were treated with

the proteolytic enzyme, Pronase, to degrade any exposed
polypeptide chains. Plasma membranes were then isolated
from these cells. These experiments indicated that approximately 10% of the plasma membrane polypeptides change
in surface exposure during development.[16,17] When SDS
gels were stained for carbohydrate with the periodic acid-
Schiff's reagent, approximately 25 major glycoproteins were
identified in the plasma membrane of vegetative amoebae and
these were completely replaced in the course of development
of vegetative cells to pseudoplasmodial cells. It is difficult to tell if any of these molecules change in surface
exposure during development since the glycoproteins of
Dictoyostelium plasma membrane are quite resistant to
Pronase treatment, either of the intact cell or the isolated membranes. If they are treated with Pronase in the
presence of detergents which disrupt the structure of the
plasma membrane, they can be digested by Pronase. The
changes in membrane composition are under developmental
control and do not result simply from starvation.[18] [19]

    Experiments described above suggested that carbohy-
drate containing components of the plasma membrane are
important in development, and initial experiments analyzing
the macromolecular composition of the plasma membrane
indicated that the most dramatic changes in its macromolecular composition occurred in its glycoproteins during
development.[16] We have developed procedures to examine
the nature and topography of plasma membrane glycoproteins
with more precision. In order to identify and analyze the
glycoproteins, we have developed a technique[15,19,20] with
which glycoproteins can be identified with great sensitivity after resolution by gel electrophoresis. Using this
procedure, gels are fixed with glutaraldehyde to cross-link
and immobilize bands of proteins. The glutaraldehyde is
washed out, free carbonyl groups are reduced by treatment by
sodium borohydride and finally fluorescent lectin is diffused into the gel in order to bind to the lectin receptor
which has been immobilized in the gel. Finally, unbound
fluorescent lectins is diffused from the gel. This procedure can be used on one-dimensional SDS or isoelectric
focusing gels or on two-dimensional gels. Its analytical
resolution is limited only by the resolution of the re-
solving gel system. Figure 2 represents an experiment
using this technique in which purified plasma membranes
from vegetative amoebae, aggregation phase cells and

pseudoplasmodium cells were analyzed by SDS gel electrophoresis. They were then stained for protein with Coomassie Blue or with fluorescent wheat germ agglutinin,
N-acetyl-chitopentose. The three gels on the left of
this figure have on them (from left to right) plasma membranes from vegetative amoebae, aggregating cells and
preculmination cells stained for protein. The three
gels in the middle present respectively vegetative amoebae,
aggregating cells and preculmination cells stained with
fluorescent wheat germ agglutinin. The three gels on the
right of this figure represent equivalent gels stained with
fluorescent wheat germ agglutinin in the presence of the
hapten inhibitor. This panel illustrates that dramatic
changes occur in the composition of wheat germ agglutinin
receptors on the plasma membrane of D. discoideum during
the development of the cell from a vegetative amoeba to a
pseudoplasmodial cell. All of the wheat germ agglutinin
receptors, except the diffuse band ranging in molecular
weight from approximately 30 to 60 kilodaltons are glycoproteins. This is indicated by their sensitivity to digestion by Pronase in the presence of SDS.[20] When similar
gels were stained with a variety of other lectins, only
receptors for the lectin Concanavalin A could be demonstrated in Dictyoselium plasma membranes.

Lectins which bind to galactose, N-acetyl-galactosamine, and fucose bound receptors in plasma membranes
that were purified from human erythrocytes. There were no
receptors for these lectins in D. discoideum plasma membranes.[20] The results of these experiments correspond in
general with what would be expected from analyses of the
sugar composition of the plasma membranes from D. discoideum.
Only mannose, glucose, glucosamine and fucose were detected
in Dictyostelium discoideum plasma membrane. No galactose,
galactosamine or sialic acid could be detected.[19] The
absence of receptors for fucose-binding lectins may indicate that the fucose presence in D. discoideum plasma membrane is present in a steric conformation which does not
allow interaction with fucose-binding lectins.

In summary, the florescent lectin-binding assay demonstrated that there are approximately 35 Concanavalin A
receptors in the plama membrane of vegetative amoebae and
that two-thirds of these change in amount during differentiation, preculmination phase cells. All of the 25 wheat

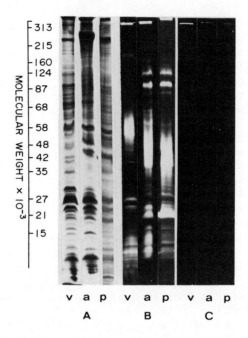

Figure 2. Wheat germ agglutinin receptors on plasma membranes of <u>Dictyostelium</u> <u>discoideum</u>. Plasma membrane proteins and glycoproteins were resolved by SDS gel electrophoresis and stained with Coomassie Blue (left 3 gels);
fluorescent wheat germ agglutnin (central 3 gels); fluorescent wheat germ agglutinin plus N-acetyl-chitopentose
(right 3 gels). Each group of 3 gels present from left
to right plasma membranes from vegetative amoebae, aggregation phase cells, and pseudoplasmodium phase cells.

germ agglutinin receptors present in plasma membrane of
vegetative amoebae disappear or are modified during the
same period of development.

TOPOGRAPHY OF LECTIN RECEPTORS

        Analysis of the topography of lectin receptors on the
plasma membrane of intact cells using microspheres conjugated to lectins, provided the following results. All
of the lectin receptors of the plasma membrane appear to
be mobile in the plane of the plasma membrane. Both

wheat germ agglutinin receptors and Concanavalin A receptors appear to uniformly cover the surface of the cell at
vegetative and aggregation stages of development. At
pseudoplasmodial stage there appear to be patches of plasma
membrane in which Concanavalin A receptors are either
absent or inaccessible for binding to the Concanvalin A
conjugated microspheres.[21]

Finally, we have used the sensitivity of the fluorescent lectin-binding assay to examine whether cells isolated from different positions along the longitudinal axis
of the pseudoplasmodium differ in the macromolecular composition of their plasma membranes. In these experiments,
pseudoplasmodia were dissected into four pieces and cells
from the anterior piece, which are destined to become stalk
cells, were compared with cells from a posterior piece,
which are destined to become spore cells. These experiments showed that five lectin receptors for wheat germ
agglutinin were present in distinctly different amounts
on the plasma membrane from the two populations of cells,
and that there were quantitative differences in the distribution of several Concanavalin A receptors. By using an
antibody specific for pseudoplasmodial antigen, we could
show that this antigen was found only in the cell from
the anterior portion of the pseudoplasmodium, those cells
destined to become stalk cells.[22]

SUMMARY

In summary, Dictyostelium discoideum provides in
microscale an illustration of the process of cellular position determination which is so important in the development
of both plants and animals. We have proposed a molecular
model for the mechanism by which this process may operate.
It has been tested in several laboratories and the results
agree with the model's predictions. Finally, it has been
possible to demonstrate that during development of such a
simple organism as the cellular slime mold, unexpectedly
substantial changes occur in the molecular composition of
its plasma membrane. These changes are most obvious in
the composition of the glycoproteins on the cell surface.
These results are in general accord with the proposition
that glycoproteins may be important determinants in cellular interaction and development.[23]

REFERENCES

1. Hunt, R. and M. Jacobson. 1974. Neuronal specificity revisited. Curr. Top. Dev. Biol. 8:203-259.
2. Bonner, J. T. 1967. The Cellular Slime Molds. Princeton University Press, Princeton, NJ.
3. Raper, K. B. 1940. Pseudoplasmodium formation and organization in Dictyostelium discoideum. J. Elisha Mitchell Sci. Soc. 56:241-282.
4. Bonner, J. T., A. D. Chiquoine, and M. Q. Kolderie. 1955. A histochemical study of differentiation in the cellular slime molds. J. Exp. Zool. 130:133-158.
5. Krivanek, J. O. and R. C. Krivanek. 1958. The histochemical localization of certain biochemcial intermediates and enzymes in the developing slime mold, Dictyostelium discoideum Raper. J. Exp. Zool. 137:89-115.
6. Jefferson, B. L. and C. L. Rutherford. 1976. A stalk-specific localization of trehalase activity in Dictyostelium discoideum. Exp. Cell Res. 103:127-134.
7. McMahon, D. and C. M. West. 1976. Transduction of Positional Information during Development in the Cell Surface in Animal Embryogenesis and Development (S. Nicolson and G. Poste, eds.). North Holland Publishing Co., Amsterdam, pp.449-493.
8. McMahon, D. 1973. A cell contact model for position determination in development. Proc. Natl. Acad. Sci. USA 70:2396-2400.
9. Pan, P., J. T. Bonner, H. J. Wedner, and C. W. Parker. 1974. Immunofluoresence evidence for the distribution of cyclic AMP in cells and cell masses of the cellular slime molds. Proc. Natl. Acad. Sci. USA 71:1623-1625.
10. Garrod, D. E. and A. M. Malkinson. 1973. Cyclic AMP, pattern formation and movement in the slime mold, Dictyostelium discoideum. Exp. Cell Res. 81:492-495.
11. Brenner, M. 1977. Cyclic AMP gradient in migrating pseudoplasmodia of the cellular slime mold Dictyostelium discoideum. J. Biol. Chem. 252:4073-4077.
12. Klaus, M. and R. P. George. 1974. Microdissection of development stages of the cellular slime mold, Dictyostelium discoideum, using a ruby laser. Dev. Biol. 39:183-188.

13. McMahon, D., M. Miller, and S. Long. 1977. The role of the plasma membrane in the development of Dictyostelium discoideum. I. Purification of the plasma membrane. Biochim. Biophys. Acta 465:224-241.
14. McMahon, D., S. Hoffman, W. Fry, and C. West. 1975. The involvement of the plasma membrane in the development of Dictyostelium discoideum. In Pattern Formation and Gene Regulation in Development (D. McMahon and C. F. Fox, eds.), W. A. Benjamin, Palo Alto, pp. 60-75.
15. West, C. M. and D. McMahon. 1977. Identification of Concanavalin A receptors and a galactose-binding protein in purified plasma membranes of D. discoideum. J. Cell Biol. 74:264-273.
16. Hoffman, S. and D. McMahon. 1977. The role of the plasma membrane in the development of Dictyostelium discoideum. II. Developmental and topographic analyses of polypeptide and glycoprotein composition. Biochim. Biophys. Acta 465:242-259.
17. Hoffman, S. and D. McMahon. 1979. The cell surface in the development of the cellular slime mold, Dictyostelium discoideum. In Surfaces of Normal and Malignant Cells (R. Hynes, ed.), Wiley, London, pp 424-547.
18. Hoffman, S. and D. McMahon. 1978. The role of the plasma membrane in the development of Dictyostelium discoideum. V. The effects of inhibition of development on plasma membrane composition and topography. Arch. Biochem. Biophys. 187:12-24.
19. Hoffman, S. and D. McMahon. 1978. Defective glycoproteins in the plasma membrane of a mutant of Dictyostelium discoideum with abnormal cellular interactions. J. Biol. Chem. 253:278-287.
20. West, C. M., D. McMahon, and R. Molday. 1978. Identification of glycoproteins and analysis of their structure using lectins as probes in the plasma membranes of Dictyostelium discoideum and human erythrocytes. J. Biol. Chem. 253:1716-1724.
21. Molday, R., R. Jaffe, and D. McMahon. 1976. Concanavalin A and wheat-germ agglutinin receptors on Dictyostelium discoideum. Their visualization by scanning electron microscopy with microspheres. J. Cell Biol. 71:314-319.

22. West, C. M. and D. McMahon. 1979. The axial distri-
     bution of plasma membrane molecules in pseudoplasmodia
     of the cellular slime mold Dictyostelium discoideum.
     Exp. Cell Res. 124:393.
23. Roseman, S. 1970. The synthesis of complex carbohy-
     drates by multiglycosyltransferase systems and their
     potential function in intercellular adhesion. Chem.
     Phys. Lipids 5:270-297.